Agents Under Fire

Agents Under Fire

Materialism and the Rationality of Science

Angus Menuge

ROWMAN & LITTLEFIELD PUBLISHERS, INC.
Lanham • Boulder • New York • Toronto • Oxford

ROWMAN & LITTLEFIELD PUBLISHERS, INC.

Published in the United States of America
by Rowman & Littlefield Publishers, Inc.
A wholly owned subsidiary of The Rowman & Littlefield Publishing Group, Inc.
4501 Forbes Boulevard, Suite 200, Lanham, Maryland 20706
www.rowmanlittlefield.com

PO Box 317
Oxford
OX2 9RU, UK

British Library Cataloguing in Publication Information Available

Library of Congress Cataloging-in-Publication Data

Menuge, Angus J. L.
Agents under fire : materialism and the rationality of science / Angus Menuge.
p. cm.
Includes bibliographical references and index.
ISBN 0-7425-3404-9 (alk. paper)
1. Religion and science. 2. Intelligent design (Teleology). 3. Reductionism. 4. Rationalism. 5. Materialism. I. Title.
BL240.3.M46 2004
215—dc22 2003020153

Printed in the United States of America

∞™ The paper used in this publication meets the minimum requirements of American National Standard for Information Sciences—Permanence of Paper for Printed Library Materials, ANSI/NISO Z39.48–1992.

In memory of Berent Enç, a fine gentleman
and a profound philosopher of agency

~

Contents

~

Illustrations

Foreword

There is a wonderful story told of an aged clergyman—a priest in the Church of England—at a meeting of the British Association for the Advancement of Science, right at the beginning of the twentieth century. There was a major controversy going on in the biology of the day, between the "biometricians," who argued that the right way to understand evolution and heredity is through the use of statistics, applied to large populations of organisms, and the "Mendelians," who adopted the new insights into (what we would now call) genetics, and who argued that the right way to understand evolution and heredity is through careful experiments on individual organisms. As so often happens in these cases, it turns out that both sides were right, and later biologists profitably used statistics and experimented on individuals. At the time, it was not so clear, and the partisans—Raphael Weldon and Karl Pearson for the biometricians, and William Bateson for the Mendelians—became heated and very personal in debate. The main point of contact and conflict was the annual meeting of the British Association (the British Ass as it was always known), and our clergyman chaired the always fiery and difficult clash. At the end of the meeting, he was expected to make a few remarks on the encounter, and he started by noting how heated the exchange had been. He then went on to refer to the projected encounter for the next year, everyone expecting him to pray for peace and friendship. Rather, with a flourish, he declaimed: "I say, this is the way to go. Fight to the death! Let the best man win, and give no quarter. The truth is more important than individuals and their reputations. Go at it!"

That man is my hero. I am not sure if he was a true Christian, but he was a true scholar. Let a thousand flowers bloom and cut down nine hundred and ninety nine for the sake of the cherished new form. Bold conjectures and rigorous refutations, as the late Sir Karl Popper used to say. To point out that the philosophical, biological, and theological positions that I take are not those of Angus Menuge is about as close to a truism as one can come in the empirical world. Menuge, a practicing Christian, thinks that modern science—particularly, modern evolutionary biology—is fundamentally wrongheaded. At least, he thinks that the philosophical propositions about reductionism and the like that underlie so much that is done and claimed in the name of modern evolutionary biology is wrongheaded. He thinks that illicit claims are made about the virtues and necessities of materialism and that there are connected false claims about the redundancy of teleology and nonnaturalistic forces and understandings. In this book, he attempts to put things right and to give both a critical analysis of those whom he thinks wrong as well as indications of the right path to be taken.

I disagree with just about everything that he says, and for that reason I urge you to read his book. Partly because I think he is wrong, and I want him refuted. Partly because he makes a good case, and he is worth refuting. A far worse sin than being wrong is being incompetent and boring—and Angus Menuge is neither of these. He makes the case for intelligent agency and against reductionism, and this should be the point for starting future debate. I note also that, strong though his convictions may be, unlike the biometricians and the Mendelians, he is polite to his opponents and careful with their arguments.

Michael Ruse
Lucyle T. Werkmeister Professor of Philosophy
Florida State University

~

Preface

Aim of the Book

This book aims to provide a rigorous defense of the intuition that scientific materialism is incoherent because it either eliminates or artificially constricts resources presupposed by materialistic scientific inquiry. In other words, scientific materialism rests on certain implicit foundational presuppositions that it is inherently unable to sustain and that are in fact incompatible with its central claims. Like so many ideas spawned by the Enlightenment, the apparent strength of scientific materialism depends on its careful concealment of the borrowed capital upon which it lives. The bankruptcy is there but undeclared. While the public image may resemble the eternally youthful Dorian Gray, the truth has more in common with his portrait.

The threatened incoherence is most clearly evident in the attempts of physicalists to explain away all, or almost all, of the appearance of design and intentionality in nature. However, it is also present in more modest attempts to naturalize the categories of agency, claiming not that these categories are illusory or nonexistent but rather that they are simply surprising aspects of the same natural world described by materialistic science. In this book, I argue in favor of the existence of significant linkages between the concepts of explanation, agency, and design that make both of these explanatory enterprises self-defeating.

The Main Argument

Perhaps the defining goal of post-Enlightenment scientific thought has been to rid science of all commitment to teleology. In particular, scientific materialism aims to show that the appearance of intelligent design in nature is illusory because undirected and automatic forces are sufficient to account for it. Thus, the "blind watchmaker" of random variation and selection has been proposed as the explanation of speciation in biology, and the algorithmic transformation of neural-activation patterns has been proposed as the explanation of cognition in psychology. An interesting question arises: Is the scientific materialist a thoroughgoing eliminative reductionist who wishes to affirm that all appearances of intelligent design in nature are illusory; or does the scientific materialist posit that while most of the natural environment is void of agency, humans and higher animals (and perhaps some aliens, if there are any) have developed by chance the real capacity for intelligent design so that some goal-directed behavior is really present in the natural world? This question turns out to be a dilemma for scientific materialism because, regardless of whether it affirms the path of the former or the latter, incoherence emerges nonetheless.

The Elimination of Agency

Suppose first that we are dealing with an eliminative reductionist who is willing to say that all appearances of design and intentionality in nature are illusory. I call this view *strong agent reductionism* (SAR). According to SAR, even the most carefully planned and rational activity of human beings, such as the development of modern digital computers, was really a complex undirected material process because goal-directed states such as beliefs and desires played no causal role. It might then be useful to adopt what Dennett calls "the intentional stance," treating human beings as if they really had goals; however, we would do so only because the stance is a convenient means of approximate prediction and because the full details of a purely materialistic account are currently unknown and would perhaps be too cumbersome if we did know them. The scientific materialist can clearly allow that treating people as if they had intentional states is useful without conceding that such states will be recognized by the ontology of a successful scientific theory of human behavior.

However, this view still leads to a number of problems. SAR is unable to account for three important phenomena: what Dennett calls the "real patterns," which emerge in human action; the existence and character of subjectivity; and the robustness of folk psychology's ontology.[1] Most decisively,

SAR also undercuts the very scientific rationality that it presupposes. At a pragmatic level, this is shown by considering what Bas van Fraassen calls the "contrastive nature of explanations."[2] Van Fraassen points out that explanations are intended to be informative. When we offer an explanation of the fact that *x* is *F*, we convey information because we explain why *x* is *F* rather than G, where G is something *x* might have been. For example, it is worth explaining why water is in solid form, because it might have been liquid. Explanations gain part of their point from the fact that the *explanandum* is not a given. Water does not simply have to be frozen—if it did, we would be much less interested in an explanation of why it was. For this reason, it is clear that the eliminative reductionist's explanations will lose their force. If everything only appears to be designed, then by default a contrast class of actually designed items does not exist; thus, it becomes, at best, much less interesting to explain why something looks designed.

But the problem is really more acute than this. If nothing is actually designed, then *design* is an illegitimate concept (a category without a "deduction" or legal title for its applicability, in Kant's terms). It then becomes a serious problem how the "intentional stance," which attributes goals and designs to an agent, can be so successful in interpreting and explaining the scientist's own behavior in constructing theories and in (as we all say) "designing" experiments. Furthermore, these appeals to an agent's designs do not seem to be dispensable because they capture high-level regularities in behavior that cannot be predicted on the basis of the heterogeneous states of the brain and body that realize them.[3]

Even more telling is the fact that the very notion of explanation employed by the scientific materialist assumes the existence of agents, that is, beings capable of directing their behavior on the basis of representations of states of affairs, such as hypotheses, predictions, plans, and designs. Agents, including scientists, make explanations in the hope that they are understood, which certainly seems to imply that it is the intention or goal of a scientist to impart this understanding. But even if the scientist's own goals could be explained away, the problem is that the act of understanding is something only an agent can do and that to explain away the act of understanding would render the whole enterprise of explanation unintelligible. The understanding of a theory consists of beliefs about what the world would be like if the theory were true. Understanding therefore requires intentionality: what is understood are possible and not necessarily actual states of affairs; therefore, such states of affairs may exhibit "intentional inexistence," in Brentano's phrase, which would subsequently disqualify any natural relation to these states of affairs, of the sort that scientific materialism would allow. Thus, there can be understanding only if there is intentionality, yet there can be explanation of any sort only if there is understanding. Therefore, the scientific explanations of the scientific materialist can succeed only if intentionality is real.

However, if intentionality is an illusion, then understanding does not exist, and neither does scientific explanation. Indeed, we would have explained explanation away. Even if the scientific materialist's explanation were "correct" and, in some Platonic sense, divorced from the actual capacities of finite intelligences, no one would be capable of understanding it; hence, things would be explained to no one. Explanation would be pointless since no one would be capable of grasping the point. The "point" of a story or explanation is precisely an intentional object, and thus it cannot remain to be grasped in a world emptied of intentionality.

Furthermore, even the claim that belief in human intentionality is a cognitive illusion seems to founder on the fact that cognitive illusions are precisely intentional states. To labor under the illusion that *p* is to falsely believe that *p*. But, of course, *p* specifies the intentional content of a propositional attitude. We are back with Descartes, pointing out that even to be deceived, one must think. But thinking is intentional, so if we are deceived, as the eliminativist reductionist claims, then intentional states do exist, in which case SAR is false. Bertrand Russell once made the perceptive point that a successful theory of truth cannot exist without an "error theory": unless you can give a plausible account of what it would be like to be mistaken, you cannot claim to have delineated truth.[4] It seems to me that, like those theories of truth that exclude the possibility of falsehood, SAR cannot be correct, because any "error theory" it proposes to account for our mistaken belief in intentionality itself presupposes that intentional states exist.

The Constriction of Goals

Suppose, however, that the scientific materialist is more moderate and allows that human goals exist and are aspects of "nature," that is, of the realm of legitimate scientific inquiry. In other words, intentionality is not to be eliminated but naturalized. I call this program *weak agent reductionism* (WAR). Prominent examples of WAR include biological and computational functionalism. According to functionalism, we do not attempt to reduce intentional states to brain states per se. Rather, intentional states are identified with something more abstract, the functional roles of certain nonintentional realizing states in mediating sensory input, subsequent states, and behavior. These functional roles may either be defined biologically, as roles that have enhanced the fitness of a creature's ancestors; or psychologically, as roles acquired by the developmental learning of the individual creature. Arguably, regardless of whether such theories pursue the biological or the psychological route, either they are faithful to materialism but fail to account for subjectivity and intentionality; or although they seem to work, they do so only by smuggling in teleological notions incompatible with materialism. In particular, it is shown in detail that Dennett's attempt to derive human intentionality from "Mother Nature" (his term for natural selection as viewed from the intentional stance) is deeply incoherent.

The repeated and systematic failures of naturalism suggest that intentionality and other characteristics of agents cannot be naturalized, at least if "nature" conforms to materialist strictures.[5] But what is the alternative? I argue that "design" and "intentionality" are legitimate but nonnatural categories. Given the fact that these categories are explanatory in the human case (which the proponent of WAR concedes), it is dogmatic to declare a priori that they must fail in alleged cases of alien, superhuman, or divine design or intentionality. Following the recent work of Del Ratzsch,[6] I argue that common arguments against invoking the nonnatural or supernatural in science are mistaken and rest, in many cases, on unexamined prejudices that derive their plausibility from an unconscious identification of empirical science with materialistic science.

Once it is conceded that designs and purposes are ever part of nature, science cannot dismiss the possibility that the realm of design extends further than humans and higher animals. Yet the contradiction present in the thinking of many opponents of intelligent design is the view that we must appeal only to material, nonintentional causes in nature, with the assumption that human scientists really have intentional states that guide their theory construction and experimental design. If humans really do have goals, as the rationality of science presupposes, then it is surely possible that other agents have goals and that we may sometimes discover empirical evidence of their activity.

Consider a simple analogy with forensics. When a forensic scientist tries to determine the cause of a person's death, we all recognize the existence of objective, evidential criteria aimed to sort out whether someone died of "natural causes," such as renal failure; because of chance, such as accidentally falling from the observation deck of Strasbourg Cathedral after an exhausting final exam; or, finally, because of the intelligent intervention of an agent. Perfectly objective reasons exist for believing that someone died because of agent intervention, even if the death was an accidental by-product of something else the agent intended to do. Being transfixed by a crossbow bolt fired from thirty paces is not the sort of thing that happens according to some periodic (repeating) regularity that could be explained by a law of nature; nor is it the sort of thing that typically happens by the chance processes of unaided nature, such as those governing quantum phenomena. Crossbows do not spontaneously generate the necessary tension to fire a bolt, even if one happens to be loaded, nor do they aim themselves at a specific site of a person so that the shot is lethal. Possibly the bolt might be fired accidentally during a struggle or while loading it for some innocuous use, but even then agent intervention is implicated. Despite the logically possible, bizarre scenarios in which chance and law might still account for the transfixion, forensic science deals not in absolute, deductive proof but with an inference to the best explanation. It can be overwhelmingly more probable that, instead of no agent being involved, a foul agent was at play or a fair agent fouled up.

The basic point of intelligent design is that now that the criteria for detecting agent activity have been made rigorous enough for serious empirical sciences—such as forensics, archaeology, cryptography, fraud investigation, and the search for extraterrestrial intelligence (SETI)—there is no a priori guarantee that those criteria will be found to apply only in the case of human or animal agency.[7] It may be that we will discover clear evidence of intelligent design in cases where no human or animal (or even alien) can credibly be implicated, such as in the complexity of biological information found in all (known) living cells or in the fine-tuning of the cosmological constants.

Finally, regardless of whether design need be invoked in understanding material events outside human provenance, the ultimate challenge for the scientific materialist is whether anything other than an intelligent designer can give a satisfactory explanation of the human capacity for design. In our own case, we have found that to give a satisfactory explanation of intentional action, we have to postulate mental states with intentional content. My argument is that the human capacity for intentional thought itself requires the postulation of a higher agency. To avoid an infinite regress, it must be supposed that this higher agency belongs to a necessary being whose intentionality is not contingent on the existence of any other agent. If intentionality is indeed sui generis, as the absurdity of reductionist materialism argues, the ultimate explanation of human intentionality is some other agent whose intentionality is self-sustaining and, hence, self-explanatory.[8]

A Road Map for the Reader

The basic structure of this book is straightforward. In the first three chapters, I prepare for and argue for, at an abstract level, the thesis of the book. In the next three chapters, I look at the specific explanations offered by scientific materialists for the apparently goal-directed phenomena in biology and psychology, and I find their arguments wanting. In the last two chapters, I defend the legitimacy of design as a scientific category against skeptical challenges, and I propose a model for the proper interrelation of science and Christianity.

Chapter 1, "Skyhooks and Cranes: The Challenge of Reductionism," prepares for the rest of the book by clarifying the varieties of reductionism and the role that they play in the program of scientific materialism. While agreeing that reductions are often major breakthroughs in science, I defend the parallel importance of nonreductive advance in science. Learning that a certain kind of phenomenon is independent of another can be at least as significant an advance as achieving a reduction of one kind to another. The focus is narrowed to reductionist accounts of agency in biology and psychology. I distinguish SAR and WAR. These are the two approaches to agency adopted by scientific materialism. SAR is committed to a thoroughgoing elimination of agency, extending even to human beings. By contrast, WAR accepts the reality of, at least, human agency but aims to naturalize it.

Chapter 2, "Strong Agent Reductionism: Materialism and the Rationality of Science," focuses on eliminative approaches to agency (especially the work of Paul Churchland). It argues that such approaches fail on their own terms and are inconsistent with the rationality of science. The scientific materialist who pursues SAR thereby loses the right to call herself a *scientific* materialist. Chapter 3, "Weak Agent Reductionism: Science and the Rationality of Materialism," considers attempts to retain intentionality as a legitimate category by naturalizing it. Standard physicalist approaches, such as Jaegwon Kim's, are shown to be inadequate. This motivates Dennett's ambitious alternative project of explaining intentionality as the product of natural selection. Dennett's account is examined in detail and is shown to be incoherent. I further argue that acknowledging the reality of agency in nature makes the in-principle exclusion of nonnatural agency ad hoc. The proponents of WAR thus lose the right to call themselves scientific materialists.

According to some philosophers and scientists, psychology will be reduced to biology. Confident that biology can be explained in purely materialistic terms, they conclude that psychology will be given a materialist reduction. However, even if psychology is reducible to biology, it does not necessarily demonstrate materialism, because it may be that biological functions themselves point to nonnatural intelligent design. This argument is the motivation for chapter 4, "Bait and Switch: Indirectness and Biological Unity," in which I consider the arguments of Michael Behe to show that certain kinds of biological structures cannot be fully explained in materialistic terms. Behe argues that biological structures that are "irreducibly complex" cannot be accounted for by the Darwinian processes available to the scientific materialist. In reply, Behe's critics have argued that Darwinian processes can generate irreducible complexity through indirect routes. I examine such proposals at length and argue that they all fail to account for the unity and cohesion of biological function. I argue that, on the contrary, top-down design is ultimately a better explanation of the facts; however, top-down design implies teleology and is therefore incompatible with any strict version of materialism.

Chapter 5, "The Alchemy of the Mind: Indirectness and Psychological Unity," parallels the previous chapter in its emphasis on functional unity and cohesion, but it focuses instead on Darwinian psychology. I argue that accounts of the self and consciousness in terms of "selfish" genes and memes[9] are unable to account for the integration and cohesion of an agent's intentional states. Indeed, the meme–gene theory of the mind is fraught with incoherence and is incompatible with what we know about psychological unity and the human capacity for practical and theoretical reason. I also argue that science itself has an indispensable commitment to the idea of agents with particular points of view and that such agents cannot be reduced to an assemblage of blind, atomic entities such as genes and memes.

In chapter 6, "Beyond Skinnerian Creatures: A Defense of the Lewis–Plantinga Argument against Evolutionary Naturalism," the case against Darwinian psychology is strengthened by a defense of arguments developed by C. S. Lewis and Alvin Plantinga. According to Plantinga, evolutionary naturalism provides no grounds for supposing that our thought processes are reliable; indeed, it tends to support the thesis that they are unreliable. Consequently, evolutionary naturalism is self-defeating in the sense that if it were true, we could never have adequate reason to believe it. Plantinga's argument echoes a distinct argument against naturalism given by C. S. Lewis in his work *Miracles*. Lewis argues that evolution operates at the level of behavioral responses that do not inherently require cognition and that the order of causation does not explain the rational order of logical thought. I develop this Lewisian argument to defend the conclusion that while reductionist forms of Darwinism might explain the development of sophisticated responses (the behavior of what Dennett calls "Skinnerian creatures"), they do not explain the existence and character of rational thought. In particular, the teleology, normativity, modal force, and personal character of practical and theoretical reasoning are not accounted for. Given this argument, the Darwinists must concede that their theory cannot explain agency or else deny agency altogether and opt for strong agent reductionism with all of the attendant problems previously noted.

Yet, eliminative skeptics will still claim that design and intentionality are fictitious categories, concepts without a deduction in Kant's sense. Against such skepticism, it is clear that empirical arguments have no force, since, regardless of how strong these arguments are, they are nonetheless compatible with design and intentionality being very useful but nonetheless fictional categories. Chapter 7, "Intentionality, Information, and Displacement: The Legitimacy of Design," develops two main arguments designed to defeat the skeptic. The first shows that the very nature of concepts implicates design and intentionality; thus, if there are any concepts, the categories of design and intentionality must be valid. However, it is possible that the eliminativist will reject the whole idea of concepts in favor of a purely information–theoretic account of cognition and action. In response, I first point out that appeal to information does not remove the need to appeal to intentionality in action explanations. But even if it did, my second argument shows that the complex specified information manifested by the theoretical reasoning and practical action of an agent still implicates design. By applying recent results of William Dembski to human cognition and action, I show that design is a demonstrably legitimate category in the human case.

Once it has been shown that design is a legitimate category, it is surely an open question how far this category extends; that is, the question of nonhuman, superhuman, or supernatural design is an empirical one, not one that can be adjudicated by a priori methodological strictures that attempt to specify in advance what science is allowed to discover. The various attempts to avoid these arguments all founder on what Dembski, in his book *No Free Lunch*, calls the "displacement problem": a common failing of naturalistic attempts to explain intentionality, design, or complex specified information by merely relocating the phenomena to be explained and so failing to show how these phenomena could have arisen from simpler kinds of phenomena.

It is a cliché of modernity that religion is the prime source of authoritarianism and dogmatism while science frees the mind from superstition, ushers in progress, and provides an enlightened perspective on the world. However, in a move that parallels the turn from rebel to tyrant in George Orwell's *Animal Farm*, some devotees of scientism, a view that claims that materialistic science is the sole source of knowledge and the only reliable guide to ontology, have become almost as doctrinaire and oppressive as the religious fanatics of the past. In chapter 8, "Science and Christianity: Dogmatism and Dialogue," I argue that the current educational policy of protecting Darwinism (and related views[10]) from serious criticism has produced a new scholasticism in science. The most dogmatic Darwinists, those who insist that scientific materialism is definitive of the scientific method,[11] use a methodology similar to the one employed by the scholastic scientists who were criticized by Francis Bacon.[12] This is what I call *uncritical deductionism*, an approach claiming to acquire "knowledge" by deducing it from preconceived ideology, rather than by serious testing against the natural world. At the same time, well-credentialed critics of Darwinism are ignored or censored and frequently suffer in their careers.

Many, though not all, of the scientists who experience such difficulties are Christians who maintain that at least some Christian presuppositions are fundamentally in conflict with the epistemology and metaphysics assumed by Darwinism. Some scholars of science and religion, such as Michael Ruse, have tried to reduce the hostility by arguing that a Darwinian can be a Christian. However, I argue that such a marriage cannot work so long as Darwinism is wedded to the reductionist stance of scientific materialism. In the last part of the chapter, I outline what I believe is a more fruitful model for dialogue between science and Christian theology, one that encourages humility and resistance on both sides. This model emphasizes the importance of a plurality of competing ideas in science and draws on the insights of John Stuart Mill and Karl Popper that concern the intellectual (and political) value of free and critical inquiry.

Acknowledgments

The research for this book was made possible by a generous fellowship from the Discovery Institute and by sabbatical leave from Concordia University, Wisconsin, both of which happened in 2002. I am very grateful to the Discovery Institute and to Concordia University for providing such a great opportunity for me to synthesize my background in the philosophy of mind with current issues in the debate between intelligent design and scientific materialism. Along the way, I have received much encouragement, advice, and constructive criticism. I would particularly like to thank William Dembski, Robert Koons, Ed Martin, Lydia McGrew, John Warwick Montgomery, Craig Parton, Del Ratzsch, Michael Ruse, Michael Saward, and all of the participants at the 2003 International Academy on Apologetics, Evangelism and Human Rights in Strasbourg, France, where the main thesis of this book was defended: Adèle Auxier, Les Barnes, Alexis Beckford, Beth and Jim Dietz, Stephen Eastwood, Hugh Gauch Jr., Susan Haden-Taylor, Sara Hoffman, Michael Job, Craig Johnson, Emmanuel Kisseh, Don and Mary Korte, Sabrina Locklair, Ronald McDonald, Dallas K. Miller, Faith St. Jean, Eric Sweitz, Angie Templer, Greg Thomas, and Steve Wood. Special thanks also go to John H. McDonald, who kindly permitted me to reproduce his diagrams of "reducibly complex" mousetraps, and to Timothy G. Standish, who generously allowed me to reproduce his diagram of the bacterial flagellum. My family should also be held up, having suffered with me loyally through the demands of this and other projects. Finally, I would like to thank Eve DeVaro and the editorial board at Rowman & Littlefield for encouraging me to bring this book to fruition.

Notes

1. Folk psychology is the common-sense, pretheoretic account of human behavior, in terms of propositional attitudes such as beliefs, desires, hopes, and fears. It assumes that we can explain an agent's action in terms of the actor's practical reasons, which are specified by these propositional attitudes.

2. See Bas C. Van Fraassen, *The Scientific Image* (Oxford: Clarendon Press, 1980), ch. 5.

3. A fascinating parallel exists between the argument for the irreducibility of mental categories in the philosophy of psychology and the argument for the irreducibility of design in the philosophy of biology. A higher-level regularity exists in both cases: in the psychological case, between agents' intentions and a class of actions with multiple physical realizations; in the biological case, between similar problems and convergent adaptations in distinct species. Because this regularity cannot be captured without postulating intentional categories, these categories are scientifically and ontologically respectable. If the psychological argument goes through, as many moderate materialists would allow, it is hard to see what besides prejudice would block the parallel biological case.

4. See Bertrand Russell, "Truth and Falsehood," chapter 12 in his *The Problems of Philosophy* (Buffalo, N.Y.: Prometheus Books, 1988), 120.

5. One can, of course, evade this problem by opting for an enriched "nature" that includes the categories of agency, but to define nature in this way would trivialize scientific materialism.

6. Del Ratzsch, *Nature, Design and Science: The Status of Design in Natural Science* (Albany: State University of New York Press, 2001).

7. Important work has been done by William A. Dembski to make the logic of the design inference explicit. See his *The Design Inference: Eliminating Chance through Small Probabilities* (Cambridge: Cambridge University Press, 1998) and *No Free Lunch: Why Specified Complexity Cannot Be Purchased without Intelligence* (Lanham, Md.: Rowman & Littlefield, 2002).

8. This argument resembles Leibniz's argument for God from the principle of sufficient reason, that the ultimate reason for contingent beings must be some being that is its own reason for existence.

9. "Memes" are discrete memorable units, such as catchphrases, which supposedly function as Darwinian self-replicators in cultural communication.

10. Many of the criticisms apply to dogmatic scientific materialists, even if they endorse a version of evolution that is not strictly Darwinian.

11. In that sense, it should be noted that Darwinism as a purely empirical, scientific proposal is not the primary problem. It is the philosophical presuppositions that require this theory to be interpreted in light of scientific materialism that create the most conflict.

12. Note that my use of Bacon's negative assessment of scholastic science in no way commits me to a naïve Baconian inductivism, derived from his positive account of "true induction."

CHAPTER ONE

Skyhooks and Cranes: The Challenge of Reductionism

> Men fall in love with particular pieces of knowledge and thoughts: either because they believe themselves to be their authors and inventors; or because they have put a great deal of labour into them. . . . If such men betake themselves to philosophy and universal speculation, they distort and corrupt them to suit their prior fancies.[1]

> The physical has been so irresistibly attractive, and has so dominated ideas of what there is, that attempts have been made to beat everything into its shape and deny the reality of anything that cannot be so reduced.[2]

Clearing the Air

Richard Dawkins complains that *reductionism* is a dirty word[3]: "Reductionism is one of those words that makes me want to reach for my revolver. . . . The only thing anybody knows about it is that it's bad."[4] The charge is unfair, according to Dawkins, because the term has been associated with a philosophy that nobody really holds. This view characterizes a "nonexistent," or "baby-eating" reductionist who "tries to explain complicated things *directly* in terms of the *smallest* parts, even, in some extreme versions of the myth, as the *sum* of the parts!"[5] Dawkins rejects such "precipice reductionism," defending a more gradual "hierarchical" approach.[6]

Hierarchical reductionism recognizes the importance of different levels of description in science because "the kinds of explanations which are suitable at higher levels in the hierarchy are quite different from the kinds of explanations which are suitable at lower levels."[7] The detailed biochemical composition of a bird's wing offers little insight into why it is so aerodynamic. Standard mechanical explanations of internal combustion do not really benefit from our measuring the spin and charm of the quarks involved. However, the hierarchical reductionist maintains that we can explain complex systems gradually, appealing to the interaction of simpler units that can themselves be understood in terms of simpler units still. In Daniel Dennett's terms, the important thing is to avoid being a "greedy reductionist" and thereby ignore the contribution of the intervening layers between the system and its simplest parts.[8] Building large structures became much easier with the development of cranes. Likewise, according to Dennett, when the complexity of a system increases, "cranes" appear, which enable novel and surprising behaviors. To use an example made famous by Michael Behe,[9] the parts of a mousetrap cannot catch mice, but the assembled mousetrap can.

Dawkins claims that this sort of hierarchical reductionism is innocuous, "just another name for an honest desire to understand how things work."[10] In the background, however, lies a controversial metaphysical assumption about what constitutes a scientific explanation. Like Dennett, Dawkins is impressed with the paradigm of engineering. From this perspective, to understand a phenomenon is to identify one or more mechanisms causally responsible for that phenomenon.[11] Indeed, that is what Dawkins means by his desire to "understand how things work." The image is of a clock or radio whose function can be understood by careful disassembly and analysis of the parts. The goal is to provide an account without appeal to nonmechanical agencies, or "skyhooks." Dennett appropriates this term to ridicule the idea of explaining something by appeal to an agency that has no mechanical support, like a hook simply hanging from the sky.[12] The value and success of the mechanistic paradigm in science is undeniable.[13] What remain controversial are the following assumptions:

1. that all phenomena can be understood in mechanistic terms;
2. that the best explanation of all mechanisms is purely materialistic; and
3. that materialistic, mechanistic explanations are the most fundamental.

One of the main aims of this book is to challenge all three of these assumptions.

Furthermore, the claim that hierarchical reductions are one and all innocuous obscures the very existence of a variety of kinds of reduction and the fact that in particular cases, one of these kinds may be vastly more plausible than the others. Further, even if all of these kinds of reduction fail, it may not necessarily be a failure for science but rather an important positive contribution, since it may establish the independence of a property or entity from others. Alchemist reductionists may have been discouraged, but it was important progress to learn that gold was an element, not a compound, and therefore irreducible to other substances. It is simply a false dichotomy to argue either successful reduction or mysterious skyhook.

To obtain a better sense of the true complexity of the issues, I begin this chapter with a survey and analysis of the varieties of successful reduction. Next, I examine the implications of failed reductions. The lessons drawn from these two discussions are then applied to the main topic of this book, attempted reductions of agency, and the focus is then narrowed to the reductionist programs of scientific materialism. The remainder of the book is a sustained argument to show there is good reason for one to reject these programs and hence to avoid becoming a scientific materialist.

The Varieties of Reduction

In a helpful essay, Paul and Patricia Churchland distinguish three kinds of reduction, all of which have had some success in the history of science.[14] I will term these kinds of reduction *conservative*, *reforming*, and *eliminative*.

Perhaps the best known of all successful scientific reductions is the demonstration that, for gases, the macroscopic term *temperature* actually refers to the mean kinetic energy (MKE) of the gas's molecules.[15] This example deserves to be called a conservative reduction because the legitimacy of the concept *temperature* is not called into question. Rather we are provided with a more fundamental reconception of just what temperature is. The reduction might have been initially disconcerting, in that the prior concept of temperature may have wrongly included as primary qualities what turned out to be secondary qualities.[16] But none of the relevant qualities are actually denied; thus, supposing (contrary to fact) that temperature were integral to cherished notions of the meaning of life, a conservative reduction of temperature need not constitute a serious threat, since both temperature and its phenomenal qualities survive the reduction intact. Since we are dealing with an identity and since identity is symmetrical, it is (for gases) just as true to say that a certain MKE is a certain temperature as it is to say that the temperature is the MKE. Although they are demoted to secondary qualities, phenomenal qualities of temperature—how it feels, for example—are not denied or proclaimed illusory. In this case, the threat of trading in what appeared to be a silk purse for a sow's ear has no merit. The mechanist reductionist is satisfied because a causal mechanism for temperature phenomena has been identified, but the antireductionist is satisfied, too, because temperature has been explained but not explained away.

The example of reducing temperature to MKE may mislead one by suggesting that conservative reductions are all mechanistic. On the pain of trivializing the thesis of mechanism, this suggestion is demonstrably not true.[17] One of the first large-scale reductions in science was the demonstration that Kepler's laws were approximately deducible from Newton's laws of motion and his law of gravitation.[18] Planetary motion, which had been thought by the early astronomers to require a special, celestial explanation, was thus identified with the same kind of motion that governs objects in the Earth's atmosphere.[19] Although Kepler was convinced that the sun exerted some sort of influence on the known planets[20] and although his three laws assume such an influence, the early astronomer's statement is purely kinematical, describing planetary motion without assigning a cause. Newtonian mechanics, by contrast, is dynamical, providing an explanation of the orbits in terms of gravitational attraction. The term *mechanics* may mislead (suggesting the paradigm of solid particles interacting like billiard balls) because the explanation is not *mechanistic*. Although Newton himself sought a mechanism for gravitation, he never found one,[21] which was precisely the initial complaint of the Cartesian mechanists who argued that gravitation was an occult force since it acted at a distance across empty space. For them, gravitation had to have an identifiable mechanical medium (a "crane"). Yet the superior predictive power and mathematical precision of Newton's theory defeated its Cartesian rivals without anyone's finding such a mechanism. While not refuting the die-hard mechanist who holds out for some future vindication, this example shows that scientific theories can be highly successful—indeed, highly successful as reducing theories—without being mechanistic.

More disconcerting than a conservative reduction is a reforming reduction, which shows that an earlier theory had significantly misconceived the phenomena it covered.[22] A classic example is the reduction of Newton's three laws of motion to the special theory of relativity (STR).[23] Newton had supposed that such quantities as mass, length, and momentum were intrinsic properties of particles, but when approximations of Newton's laws are deduced from STR, we learn that in fact these quantities are relations that cannot be fixed without appeal to a particular frame of reference. Newton supposed that a particle p simply has mass m; however, STR shows that while p may have m relative to frame of reference F, p may also have a different mass m^* relative to an alternate frame of reference F^*.

Most would agree it is going too far to say that STR shows that Newtonian mass simply does not exist. The Newtonian conception of mass is significantly mistaken, but the amount of misconception is consistent with successful reference to something real.[24] Nonetheless, had the meaning of life depended on Newtonian mass, observers would feel a sense of disorientation and the need, at the least, for a period of readjustment. There is violence to common sense—it is obvious to "folk physics"[25] that a body has one particular mass—and already part of what was believed has not merely been explained but explained away. STR explains that bodies seem to have just one unproblematic mass because we always view them from a particular frame of reference. Our belief in intrinsic mass is an illusion of our parochial situatedness, yet mass lives on. We have learned to think in clearer and more precise terms about what we previously glimpsed through a mist. But what we glimpsed was real.

Most radical of all is eliminative reduction. Here, the old conception is shown to be so wide of the mark that it just needs to be thrown out and replaced. The old idea is not modified but deleted. For example, no single gas has all the properties attributed to phlogiston,[26] and phlogiston theory has not been reformed but replaced by a theory that distinguishes oxygen, carbon monoxide, and carbon dioxide. Country singers and journalists still allude to centrifugal force, but there is no such thing.[27] For that matter there is no caloric fluid, there are no humors (a melancholy thought), and there is no aether.

If we suppose that the meaning of life hinged on phlogiston, we would likely suffer from vertigo. Consider, for example, human rationality. To learn that our conception of this is flawed by excessive idealization of our fallible capacities is unsettling but tolerable. However, if we were to learn that, when properly understood, no such thing as human rationality existed, the bottom would drop out of our world; and it would have vast repercussions and the potential for outright incoherence and nihilism. We may be convinced that some old categories for understanding the physical world need to be thrown out, however long they had been cherished in our attic. But we are understandably more than a little reluctant to throw out as trash what seems to be a significant part of ourselves, a part without which life seems quite pointless.

To see that this response is not merely reactionary, we need to consider what the point of reductionism itself is. Reductionists aim to remove mystery, substituting cranes for skyhooks. But certain kinds of purported reduction increase mystery because the proposed crane makes so much of what was previously understood incomprehensible. The reduction of Kepler's laws to Newtonian mechanics, and of this to STR (and the properties of spacetime), increases our understanding of the phenomena. The replacement of human rationality with some alien category might eliminate a troublesome skyhook but only by making human action, including science itself—as a preeminent example of rational human activity—inscrutable. I will develop this point more fully in the next chapter. The important point for now is that it is simply false to claim that reductions are to be valued simply because they rid us of something we cannot explain. The test of a successful reduction is not whether one more thing can be explained (the crane that is substituted for the skyhook) but whether there is a net increase in explanatory power, which won't happen if the crane mystifies what the skyhook made plain.

When elimination seems threatening, the ardent reductionist often tries damage control measures. One of these is to claim that the replacement category is really not all that bad and that once we have jumped ship, we will get used to it. This scenario may be plausible if the new category actually saves the phenomena that the old category succeeded in capturing. Phlogiston theory, for all its faults and need for ad hoc assumptions, actually did quite a good job of predicting the experimental results and what phlogiston theory was right about is not lost by moving to oxygen theory; thus, the replacement is a success. Again, the reductionist may argue that we do not need to save certain phenomena, only the appearances, because the phenomena (as we had thought of them) were illusory. This is what Darwin did, arguing that biological structures only seemed to be designed. The phenomena of designed artifacts in nature were not saved but proclaimed illusory; however, Darwin claimed that he did still save (explain) the appearance of design. Nonetheless, if the reductionists offer a replacement category that fails to save even the appearances, then they are engaging in a confidence trick, a bait and switch. The bait is losing a skyhook; the switch is losing a plausible account of what the skyhook explained. As we shall see, proponents of intelligent design make just this case against Darwinism, identifying the appearance of design with the manifestation of a certain kind of information (complex specified information) and arguing that natural selection is unable to generate it.

Another damage control method is to finesse the distinction between reforming and eliminative reductions. Since reform is less threatening than elimination, it is tempting to dress up an execution as substantial correctional measures. This scenario is sometimes plausible because the vagueness of our concepts makes the distinction between the two kinds of reduction hard to draw. For example, because computation can be conceptualized in many ways, it can be quite unclear what it means to reduce cognition to computation. Such a reduction may seem to rob human thought of its intentionality because, on the face of it, the computer's states are not really about anything. But the reductionist may offer an ersatz version of intentionality that captures at least some of what we believed we had before. How much has to be retained for the reduction to be merely reforming rather than eliminative? Attempts by antireductionists to save the phenomena may be blocked by the reductionist's claim that the phenomena are illusory. Many philosophers argue that humans have original intentionality[28] (intentionality that is not, like the water in our body, derived from another source). However, Dennett maintains that original intentionality is an illusion because our intentionality is in fact derived from "Mother Nature."[29] He claims that intentionality is dependent on the notions of biological and psychological function, which can only be defined in evolutionary terms. Dennett presents himself as offering a reforming reduction of human intentionality—intentionality does exist, but it is different than we thought—whereas others see him as an eliminative reductionist. And although the same word is being used, what Dennett means by *intentionality* is as different from our prior conception of intentionality as oxygen is from phlogiston.

Crosscutting the threefold distinction between kinds of reduction is a twofold distinction between modes of reduction. These modes are usually termed *synchronic* and *diachronic*. I will continue to use these terms, but I would note that they are unfortunate in drawing the distinction in strictly temporal terms: in the synchronic mode, a property or entity is reduced to some simultaneously present property or entity; in diachronic mode, there is a reduction to the characteristics of a causal ancestor of the system being studied. Strict simultaneity is, however, surely not the essence.[30] For example, if prior brain states causally determine mental states, then the reduction should be considered synchronic because the brain states are intrinsic to the system. However, if a causal theory of representation locates the origin of content in a creature's historical relation to its environment, then the reduction is certainly diachronic because the account depends on features extrinsic to the system. As I shall use the terms then, synchronic reductions show how a feature derives from intrinsic characteristics of a system whereas diachronic reductions show how the feature derives from the extrinsic causal interactions with the system's environment. Thus, the following reductions would all be synchronic:

- the temperature of a gas to the MKE of its molecules, a conservative reduction;
- "folk psychology" to functionalism, a reforming reduction (discussed later in the chapter); and
- phlogiston to oxygen, an eliminative reduction.

But the reduction of Kepler's laws to Newtonian mechanics is a conservative diachronic reduction because it reduces planetary motion to the action of gravitational causes. Likewise Dennett's proposed reduction of human intentionality to Mother Nature, via a reduction of biological and psychological function, is either a reforming or an eliminative diachronic reduction.

Failed Reductions and Nonreductive Advance in Science

Reductionists often use rhetoric to suggest that resistance to their favored reductions is obscurantism. The opponents of reduction are portrayed as reactionary holdouts, afraid of progress and clinging for comfort to tradition or superstition. However, this bullying technique falsifies the way science has actually progressed.

Failed reductions, including the failure to provide a mechanism, do not necessarily mean a failure for science. Nor need they even mean that science is treading water. They can actually signify substantial progress and an increase in our knowledge.

Alchemists were diachronic reductionists[31] who believed it was possible to generate precious metals, such as gold and silver, by transmutation of base metals. The plausible mechanistic assumption was that gold and silver could be obtained by reorganizing other materials. The idea was partially true, since one can generate compounds in this way. But it was falsified by the discovery that gold and silver are discrete elements with unique atomic structures. Establishing the independence of the elements from one another was surely the most important discovery in launching the modern science of chemistry, yet that progress involved the refutation of a highly appealing reductionist research program.[32] Furthermore, in repudiating alchemy, modern chemistry was not arguing from ignorant obscurantism but from greater knowledge. It was able to explain why alchemy was doomed in principle. It did not merely assert, on the grounds of its being so precious to us, that gold was sui generis. It provided a compelling explanation of why this was the case. At this point, the reductionist who still clung to alchemy would be the obscurantist. The mechanist intuition that we can always synthesize an entity from something else is not always vindicated by science.

The success of the conservative Newtonian reduction of Kepler's laws is not impugned by the fact that Newton did not offer a mechanism. The Cartesian mechanists had hoped to offer a universal reduction of motion based on the direct contact of particles. Since gravitation had no such mechanical basis, the ascendance of Newtonianism was a failure for Cartesian reductionism. Yet there is no question that the Newtonian paradigm was a huge advance in science, unifying the previously disjoint realms of celestial and terrestrial physics and allowing an unsurpassed accuracy of prediction of the orbits of planets and even the "erratic" comets.

Again, an attempt was made to reduce the propagation of electromagnetic (EM) waves, described by James Maxwell's theory, to the mechanical properties of aether. However, the attempt failed.

> Unexpectedly, the existence of such an absolute medium of luminous propagation turned out to be flatly inconsistent with the character of space and time as described in Einstein's 1905 special theory of relativity. EM theory thus emerged as a fundamental theory in its own right, and not just as a special case of mechanics. The attempt at subsumption was a failure.[33]

In this case, as in the others, scientific progress is made through a failed reduction, by learning that certain phenomena are independent of others. STR was another enormous advance in science made in spite of, indeed partly because of, its thwarting of a mechanistic reduction.

In each of these three examples, qualities or entities resist reduction to a mechanical base, yet they not only retain but increase explanatory power. Burning down the cranes erected to explain the origin of the chemical elements and the transmission of gravitation and electromagnetism did not prevent these ideas from vastly increasing the explanatory power and range of science. If the opposite of a crane is a skyhook, then skyhooks are frequently good for science. In fact I suspect that Dennett's aversion is really to supernatural skyhooks. Yet since both natural and supernatural agencies can fail to be cranes, the lack of a causal mechanism provides no reason to accept one and reject the other. To do so would require another argument, one in favor of natural skyhooks and against supernatural ones.

Successful reductions can be nonmechanistic. Mechanism and reduction do not have an inherent connection.

As we have already noted, Newton's conservative reduction of Kepler's laws did not provide a mechanism for planetary motion but an unanalyzed force. Again, the partial reduction of chemistry to quantum mechanics is a case where a paradigmatically mechanistic theory of the formation of chemical compounds is itself reduced to a theory that is significantly nonmechanistic. As in the Newtonian case, "mechanics" misleads. Mechanism, in the sense characterized by cranes, implies the existence of a structure consisting of parts that occupy determinate locations and operate continuously through physical intermediaries that are localized in space and time. None of this is evident in the quantum world. For example, consider the individual electrons within an atom. According to the Heisenberg uncertainty principle, it is not possible to simultaneously fix the location and momentum of an electron; therefore, one or both of the concepts of location and momentum may not really apply to such particles in the same way as they do to macroscopic phenomena (indeterminacy). Furthermore, electrons jump between valences without a smooth transition (discontinuity); in fact, some quantum phenomena even violate locality, with independent events appearing to influence each other at more than the speed of light, a mechanical impossibility (nonlocality).[34] Quantum mechanics is probably the best-confirmed scientific theory to date, yet it invokes powers that resist any classically mechanical reduction or interpretation.[35] Quantum mechanics is as full of "skyhooks" as it is of explanatory and even reductive power.

A scientific materialist would undoubtedly concede the legitimacy of "skyhooks" in quantum mechanics because they still seem to be parts of the natural world. However, if violations of mechanistic strictures are compatible with scientific materialism, we are owed an account of why supernatural violations are more problematic than natural ones. If even materialistic science leads to the postulation of skyhooks, the question should be "Why are supernatural skyhooks more problematic than natural ones?"

Reductionist zeal can be bad for science. Reductionist enthusiasm for simplification can obscure and falsify important phenomena, thereby inhibiting scientific understanding.

Although they are eliminative reductionists with regard to the mind, the Churchlands readily concede that the reductionist impulse can be misplaced, sometimes because the reducing theory does not really save the appearances: "It is of course a bad thing to try to force a well-functioning old theory into a Procrustean bed, to try to effect a reduction where the aspirant reducing theory lacks the resources to do reconstructive justice to the target old theory."[36]

Nonetheless, such attempts are commonplace in science. Bacon complained that Aristotle and his followers had attempted to force all phenomena into the mold of preconceived essences, as if nature were obliged to obey the expectations of human reason. And "Gilbert too, after his strenuous researches on the magnet, immediately concocted a philosophy in conformity with the thing that had the dominating influence over him."[37] Bacon's point is that we become impressed with the power of a particular paradigm or model, so we start to conceptualize everything in those terms. To do so may, of course, be fruitful. In fact, for both Kepler and Newton, magnetism was a helpful analogy in the formulation of their theories. Yet Bacon warns that the human mind is always apt to impose preconceived order on recalcitrant appearances. Gravitation turned out to be quite independent from magnetism.

Today, like the British "chippy," a shop that serves chips with everything, computational and evolutionary theories are proposed about anything under the sun. As we shall see, it is quite clear that some of these theories shed more darkness than light, at least for those not in their grip. A kind of

Hegelian fallacy[38] supposes that science has at last made a grand synthesis, or "consilience," when the success of a model in one domain has led to unwarranted extensions to quite different phenomena. In the attempt to impose this model everywhere, important phenomena are treated like political dissidents under communism, either warped and disfigured by forced assimilation or silenced by exile and execution. As Thomas Nagel has emphasized, such distortion and suppression of the truth is encouraged by a prevailing attitude of scientism, not science itself but rather a certain attitude of veneration toward science as the exclusive means of obtaining truth:

> Scientism is actually a special form of idealism, for it puts one type of human understanding in charge of the universe and what can be said about it.[39] At its most myopic it assumes that everything there is must be understandable by the employment of scientific theories like those we have developed to date—physics and evolutionary biology are the current paradigms—as if the present age were not just another in the series.[40]

As I argue in chapter 5, evolutionary psychology provides many instances of this tendency. Evolutionary psychologists have made claims that rest on an unwarranted extension of evolutionary theory, which conflict with what psychologists and philosophers already know about the mind;[41] which postulate hypothetical entities that do not account for such crucial aspects of the mind as subjectivity, psychological unity, or practical reason; and for which there is no independent evidence.[42]

Worse, it is extreme reductionists who are most likely to adopt the ploy of identifying their favored reductionist program with "science" or "the scientific method." From this perspective, critics of the program are dismissed as antiscience obscurantists. The tactic is as scurrilous as it is indefensible. As we have seen, many reductions fail, and the progress of science does not exclusively depend on the success of reduction, and it depends far less on the success of the particular reduction favored by the extremist. Science can progress when reductions fail and even because reductions fail, so it is nonsense to suggest that opposition to reduction is necessarily antiscience. When reductionists dismiss their critics in this way, it is understandable that reductionism is treated as a dirty word.[43]

Agent Reductionism

Having surveyed the geography of success and failure in reduction, our focus narrows to the main concern of this book, attempted reductions of agency. For this section, I will first define the relevant notion of agency, then I will explicate the possible varieties of agent reductionism.

Agency is sometimes understood in senses that are so metaphorical and anthropomorphic that one can even talk of a chemical-cleaning agent without supposing for a moment that this agent has any beliefs or desires. Here agency will be understood in the fullest, most literal sense as applying to anything that has representations of its own goals, such as desires, and the means to achieve those goals, such as beliefs, and whose behavior is rendered intelligible in light of those representations. It is not enough that the entity represents goals or means but that these must also, in virtue of their mode of representation, be goals or means *of* the entity in the sense that they provide the entity's own reasons for acting. As an agent in this sense, when I want a beer and believe there is one in the refrigerator, it is significant that I do not merely represent beer and opening the refrigerator but that I represent *my* having a beer by *my* opening the refrigerator. This means that the representations belong to, rather than merely occur in, the entity and that the entity is not a passive spectator of the unfolding of its actions but their author. It also means that a true agent must be capable of self-representation, regardless of whether it is "conscious" in some more specific sense.[44]

A defining characteristic of all these representations is intentionality. That is,

- the representations are about something (they have a potential reference);
- they characterize that thing in a certain way (they have a content);
- what they are about need not exist (reference can fail);
- where reference succeeds, the content may be false; and
- the content defines an intensional context in which substitution of equivalents typically fails.

For example, one can believe that Santa has a white beard, even if there is no Santa or, if there is, even if his beard is black. And one can believe that Santa has a white beard without believing that his beard falls within the wavelengths of light that cause the phenomenological property of white. Thus, it goes without saying that where there is agency, there is intentionality.[45]

The notion of agency in this sense is suggested by behavior or structures whenever they appear not to be the result either of chance or of natural laws operative in the environment.[46] Neither chance nor any known law (such as one governing magnetism) explains my opening the refrigerator, but we do understand this action as one issued from my own agency. By contrast, some, such as Paul Churchland, believe that the idea of agency derives from a false folk psychology, which will one day be displaced by a developed neuroscience. Again, artifacts suggest agency. On the basis of the empirical fact that complex biological structures are much like human artifacts in their complexity and in the mutual adaptation of the parts subservient to a common function,[47] the British natural theologians (including William Paley and Thomas Reid) argued that the agency of a supernatural designer, not law or chance, was the best explanation of these structures.[48] By contrast, Darwin argued that we were witnessing only the appearance of design, which could be explained by the interplay of chance variation and lawful natural selection.

It is instructive to consider how the kinds of reductionism may be applied to debates in the philosophy of mind about human agency and to the spectrum of positions in the creation–evolution debate. For instance, in the philosophy of mind, identity theorists[49] are clearly acting as conservative reductionists when they claim that all the mental properties associated with agency are type identical with abstract physical properties of the brain. It is actually much less clear than usually admitted whether it is significant for the identity theorist to claim to be a physicalist. If the identity succeeds in capturing the full notion of agency, then the mentalist has lost nothing valued; therefore, the insistence that all reality is "physical" seems either arbitrary—that is, if mental states are identical with physical ones, then it is equally true that the physical states are mental—or dependent on a suspiciously inflated notion of what counts as physical, which simply absorbs problematic mental characteristics.[50]

Identity theories fell out of favor for several reasons. One, pressed by Donald Davidson, was that the ascription of mental states is governed by norms of rationality that have no counterpart in physical science.[51] This argument suggested that it was unreasonable to expect the kind of strict psychophysical correlation required for an identity. Davidson proposed instead a mere token identity between mental and physical events and that physicalism must be weakened to supervenience. According to supervenience, mental states are completely determined by their physical bases, but the same mental state may have multiple bases.[52] This argument suggested another idea, functionalism,[53] according to which mental states are functional roles that can be "realized" by different physical implementations.

Functionalists typically deny that they are reductionists because they do not believe mental states are reducible to the physical states of a particular kind of substrate, such as the neuronal structures of the human brain. The abstract function of a mousetrap can be "multiply realized" by a variety of physically different mouse-catching mechanisms. Likewise, perhaps the functions characterizing cognition could be realized in substrates quite different from the human brain (computers, aliens, and other animals might be examples). However, what the functionalist is rejecting is the type reduction of identity theorists, not reductionism per se.[54] Multiple realization does not preclude all forms of reduction. Temperature is realized quite differently in gases, solids, and vacuums; but this is no argument against the possibility of reducing temperature, only against a uniform, domain-indifferent way of doing so. Nor would it prevent a functionalist from claiming that some common function of temperature could be achieved in all of these realizations. In fact, functionalism can be understood as an attempt to reduce the folk-psychological conception of mental states to the (allegedly) more scientific notion of functional roles that can be implemented by physical systems. This is perhaps best understood as a reforming reduction, attempting to retain folk psychology's commitment to intentionality (via the notion of function) while providing a superior account of the causal explanation of behavior (via the function's implementation).[55]

Reinforcing a point made by Jerry Fodor, I should note that both identity theorists and functionalists are synchronic reductionists whose programs (if correct) would remain intact regardless of the fate of an attempted diachronic reduction, such as that proposed by Dennett.[56] Dennett's position, the view that human intentionality is derived from Mother Nature, is therefore not in the least obvious, even to those wedded to some form of reductionism in the philosophy of mind. Thus, even if reductionism were the only alternative to obscurantism (which it is not), no one should feel pressure to accept Dennett's position as the only reasonable option. Dennett's position is also hard to place because it is not clear whether he is reforming our notion of agency or actually replacing it with a new notion that bears only a metaphorical relation to agency as we have defined it here.

By contrast, Paul Churchland proposes a seemingly straightforward eliminative reduction—eliminative materialism—which requires a radical reconceptualization of all rational human activity. Churchland maintains that folk psychology (FP) is just like folk physics, a prescientific but nonetheless empirical theory that fails to do justice to the empirical facts. According to FP, propositions and inferences play a role in human cognition, but neuroscience finds no empirical support for this assumption, as all that can be observed are computational transformations of patterns of neuronal activation, or "activation vectors." "The brain appears to be playing a different game from the game that FP ascribes to it."[57] Since the representations count as reasons because of their propositional content, Churchland's reduction would certainly undermine a traditional understanding of practical reason. Eliminative materialism implies, quite simply, that there are no such things as beliefs and desires, at least as FP thinks of them. Whether eliminative materialism could nonetheless be squared with a highly reformed but still recognizable form of agency would depend crucially on whether intentionality can survive in a world of activation vectors. Churchland's view will be critically evaluated in chapter 2.

Emerging from the creation–evolution debate are four main contenders with many variants, epicycles, and intermediate blends: Darwinism, self-organization, theistic evolution, and intelligent design. For the sake of brevity, only the first and last of these will be considered at any length.

According to Darwinism, design in biology is not a genuine phenomenon but an illusion to be explained away. For Darwinists:

> Biology is the study of complicated things that give the appearance of having been designed for a purpose. . . . Natural Selection, the blind, unconscious, automatic process that Darwin discovered, . . . the explanation for the existence and apparent purposeful form of all life, has no purpose in mind. . . . If it can be said to play the role of watchmaker in nature, it is the *blind* watchmaker.[58]

Thus Darwinism argues that the *explananda* are not what the British natural theologians thought: they are not artifacts but only structures like artifacts. This might seem like bait and switch, but, according to Dennett, "evolutionary bait-and-switch is not really nefarious; it just seems to be, because it doesn't explain what at first you thought you wanted explained. It subtly changes the topic."[59] On this reading, Darwinism thus proposes a diachronic eliminative reduction of biological design.

Where Darwinists disagree with one another is on what survives this reduction. Skeptics argue that even the idea of biological function lingers on only as a useful fiction. The fiction is useful because it facilitates reverse engineering, the attempt to figure out how a complex structure is built.[60] It is practically impossible to do this without treating the structure as having some particular function. Yet if, as a skeptic may maintain, only that which is designed for a purpose can literally have a function, the ascription of function is at best metaphorical. Others attempt to retain the idea of function by claiming that there is such a thing as selection *for* certain features.[61] Thus, it may be claimed that the heart is selected for pumping blood and not for producing a certain rhythmic sound, which is why the heart really does have the function of pumping blood but not of making that sound. The obvious problem with the proposal is that the ability to select for a process appears to imply foresight and even agency, precisely those characteristics that Darwinism denies can be ascribed to natural selection. Yet, if natural selection is only metaphorically treated as an agent, it is hard to avoid the skeptic's charge that function ascriptions are likewise metaphorical.[62] However, as Fodor points out, the failure of a diachronic account to capture the idea of function does not preclude a successful synchronic account: "My intuition . . . is that my heart's function has less to do with its evolutionary origins than with the current truth of such counterfactuals as that if it were to stop pumping my blood, I'd be dead."[63]

Likewise, Darwinists are divided on the issues of human psychological function and intentionality. Many of those committed to "evolutionary psychology," such as Dawkins and Dennett, admit that a purely biological account of cognition based on our genes is inadequate. However, they propose to supplement the effect of our genes with an account of "memes," the ideas and linguistic structures of cultural evolution. An evolutionary account of the origin of genes and memes would provide a diachronic reduction of the content of human cognition, although Dawkins admits that it really does not obviously account for such phenomena as consciousness.[64] For such thinkers, the reference to memes is not an admission of the limitations of Darwinism but an extension of it. Memes, like genes, are understood as Darwinian replicators, that is, units capable of mutation, selection, and differential reproduction.[65]

However, Fodor is right to emphasize that being a Darwinist about biology does not require one to be a Darwinist about the mind, since "we could know the whole story about how the mind supervenes on the brain, without knowing *anything* about the evolution of either."[66] Furthermore, Darwinists such as Richard Lewontin and Kenneth Miller[67] argue that when creatures are capable of high levels of reason, as in the case of human beings, their thought and behavior becomes largely independent of evolutionary considerations. As we shall see later in the book (especially chapters 5 and 6), it is highly dubious that a satisfactory account of such rationality can be constructed on the basis of genes and memes alone.

Briefly, self-organization (SO), while not a flat denial of Darwinism, argues that the mechanism of natural selection is incapable of fully explaining the formation and stability of complex systems.[68] It argues that the inherent properties of matter are akin to the intrinsic tendency of minerals to form crystals with a particular geometry and collective organization, which would account for the development and perseverance of complex, delicately balanced biological structures. Nonetheless, the outworking of these inherent tendencies is assumed to be lawlike, not requiring any invocation of design. Thus, SO, like Darwinism, aims at a diachronic eliminative reduction of design in biology.

Proponents of theistic evolution (TE) may partially accept the insights of Darwinists and self-organizers, but they would insist that none of these processes would have been successful in producing life and its diversity without the guidance of God. This guidance might take the form of ongoing intervention or merely the fine-tuning of the initial conditions of the universe. In either case, should the theistic evolutionists maintain that God's guidance was detectable as design, they would also count as proponents of intelligent design (see below). Yet, some proponents of TE claim that God has aided the evolutionary process in ways inaccessible to empirical science. A theistic evolutionist such as Howard Van Till, who argues that we can trace life back to the "in-built potentiality of creation" or a "robust formational economy," can be a diachronic reductionist relative to the universe's initial conditions while denying that the conditions themselves have such a reduction. However, an interventionist proponent of TE would deny that diachronic reductions to initial conditions can be successful because God superadds characteristics to creation so that its subsequent development cannot be reconstructed by appeal to the resources available before that intervention.[69]

The latest contestant in the debate is intelligent design (ID), which argues that design can be understood as sui generis, irreducible to chance, law, or any combination of the two.[70] In that sense, ID is plausibly interpreted as a nonreductionist research program, and as we have seen, it would be no less scientifically respectable for that. If ID is successful in showing that design is not reducible to the effects of laws and chance alone, this would be at least as important a scientific discovery as Einstein's demonstration of the independence of the electromagnetic phenomena described by Maxwell from classical, mechanistic categories.

There is, however, an area of potential confusion here.[71] In the ID literature, some references to "design" are not to design as a cause (detectable or not) but to design as an empirically detectable effect. It is important that these two senses of design are carefully distinguished. The epistemic sense of design (detectable effect) is much more restricted than the ontological sense (cause). Some genuine design may not leave an empirically detectable trace. A clever murderer may mimic accidental or natural death, thereby making a malicious design undetectable. As an empirical, epistemic concept, design must be restricted to those cases where chance and law can be reliably excluded. Nonetheless, design may be at work incognito even when chance and law cannot be ruled out as explanations. One suggestion is that design as an empirical effect can be identified with the manifestation of a certain kind of information, complex specified information (CSI), which is the idea behind the explanatory filter proposed by William Dembksi.[72] Design may safely be inferred only if a behavior or structure is contingent, complex, and conformable to an independent pattern or specification. If contingency fails, a law of nature might have produced it; if complexity is lacking, it could be chance; and if an independent pattern is lacking, it might still be chance. However, if the entity is contingent, complex, and conforms to an independent pattern, then by definition it exhibits complex specified information: "To infer design by means of the complexity-specification criterion . . . is equivalent to detecting complex specified information."[73]

Many proponents of ID argue that design in this sense is empirically detectable in cases where Darwinists claim that law and chance suffice. The argument is that biological structures exhibit CSI but that the Darwinian mutation–selection mechanism (with or without the supplement of self-organization) is incapable of generating CSI. Furthermore, ID is able to reconcile a fully robust notion of human and divine agency with its picture of science.[74]

Agent Reductionism and Scientific Materialism

With a survey of the varieties of agent reductionism behind us, it is now important to define two particular stances that provide the pivotal focus for the rest of the book: *strong agent reductionism* (SAR) and *weak agent reductionism* (WAR).

The SAR denies the ultimate legitimacy of the notion of agency (as defined in this chapter), even if it is somewhat reformed and even in the case of human beings. The strong agent reductionist is thus an eliminativist (synchronic or diachronic), even with regard to human intentionality and practical reason. Folk psychology is not somehow constitutive of humanity but merely an internalized false empirical theory awaiting a replacement theory, whether framed in the terms of neuroscience, the interaction of genes and memes, both, or neither.

The WAR claims that although evolution is void of agency, human beings do have the real thing. There is thus no eliminative diachronic reduction of human intentionality and practical reason to natural selection, nor is there any synchronic eliminative reduction of the kind proposed by eliminative materialism. The weak agent reductionist is an eliminative reductionist with regard to the apparent agency of natural selection; but with regard to human intentionality, the weak agent reductionist is either a nonreductionist or a conservative or reforming reductionist. The proponent of WAR may allow the possibility of agency in some nonhuman animals or unknown aliens but denies that it played any primary[75] role in biological origins.

Scientific materialism is the view that legitimate scientific explanations invoke only materialistic categories. It seems to me that any scientific materialist worthy of the name must embrace either SAR or WAR. Either agency as we have defined it is incompatible with scientific materialism or it is not. If it is, the scientific materialists should embrace SAR; if it is not, they should embrace WAR.

In the next two chapters, I will outline general arguments to show that neither SAR nor WAR is tenable. SAR can be shown to be both internally incoherent and incompatible with the activity of science itself because robust notions of practical reason and intentionality are constitutive of science but denied by SAR. Attempts to eliminate these notions demonstrably fail. Not only that, the endorsement of such attempts is incompatible with the rational practice of science. The proponents of SAR thus lose their right to proclaim themselves scientific materialists. However, if WAR is adopted, the admission that agency is a scientific reality in the case of human beings makes the in-principle exclusion of the possibility of nonmaterial (disembodied and/or supernatural) agency ad hoc. If any agency is real and detectable—and the detectability of the effects of agency does not depend on the character or detectability of the agent—then the question of whether nonmaterial agencies are operative in nature is an empirical question. This does not show that scientific materialism is false but that attempts to dismiss ID as nonscience, on the a priori ground that science cannot allow nonmaterial agencies, fail. However, what may show that even WAR is false is a demonstration that the capacities of human agency cannot be given a materialistic explanation because these capacities are contingent on another, nonmaterial agency. If this is correct, the proponents of WAR may be unable to proclaim themselves scientific materialists. In either case, scientific materialism fails. Either the scientific materialist poisons the rationality of science or opens the door to the nonmaterial; in either event, the proponent is no longer a scientific materialist. If I am correct that scientific materialism has no alternative other than SAR and WAR, it follows that scientific materialism is false.

Predictably, some will construe this as an attack on science because they identify scientific materialism with "the scientific attitude." My response is that scientific materialism is neither an implication nor a presupposition of doing science. Rather, it is a metaphysical and sometimes religious stance that some people have toward science. In its strong form, that of SAR, the stance is actually incompatible with the rationality of science. In its weak form, that of WAR, the stance excludes the possibility of certain types of explanation on the basis of an arbitrary prejudice against the nonmaterial, backed by the conceptual inertia of Enlightenment and positivist paradigms of scientific knowledge.

Conclusion

This chapter has involved a lot of survey and definition to set the stage for the rest of the book. Here is a summary of its main points.

1. Greedy or precipice reductionists are largely a figment. Serious reductionists take a gradual or hierarchical approach, using the cranes supplied by the intervening layers and acknowledging the need for multiple levels of explanation.
2. Successful reductions are of significantly different kinds and modes. There is no such monolithic thing as "the reductionist attitude." One can be a reductionist on some issues and not on others. The failure of one kind or mode of reduction does not preclude the success of another kind or mode of reduction. In particular, the diachronic eliminative approach to reduction favored by Darwinists is not the only style of reduction in town.
3. The failure of particular reductions or of reduction in general is not necessarily a failure for science. On the contrary, failed reductions may allow great scientific advances. It is an important scientific insight to learn that a property or entity is independent from others.
4. The failure of mechanistic reductions does not necessarily impede science or even preclude nonmechanistic reductions. The dichotomy "either unscientific skyhook or scientific crane" is false because nonmechanical skyhooks can be very useful in science and mechanical cranes can be ideological figments. Mysterious skyhooks may shed a great deal of light, unifying and reducing other phenomena, whereas cranes may cast the whole world into darkness by denying the facts in front of our face.
5. Botched or overblown reductions can hinder science by obscuring realities that need to be explained. The identification of science with a favored reductionist program is a deplorable kind of scientism that prohibits the atmosphere of free inquiry and honest debate essential to true science.
6. The notion of agency is the crucial target of reductionist programs in biology and psychology. Proponents of SAR treat all agency, even human agency, as illusory; proponents of WAR accept human agency but reject agency elsewhere, with the possible exception of higher animals and aliens. Scientific materialists are committed to either SAR or WAR. The next two chapters develop the main reasons for rejecting scientific materialism on the grounds that neither SAR nor WAR is tenable.

Notes

1. Francis Bacon, *The New Organon*, ed. Lisa Jardine and Michael Silverthorne (Cambridge: Cambridge University Press, 2000), bk. 1, 54: 46.

2. Thomas Nagel, *The View from Nowhere* (New York: Oxford University Press, 1986), 15.

3. Richard Dawkins, *The Extended Phenotype: The Gene as the Unit of Selection* (Oxford and San Francisco: Freeman, 1982), 113.

4. Richard Dawkins, in an interview with Steven Pinker, "Is Science Killing the Soul?" *Edge* 53, April 8, 1999, 15, available online at www.edge.org/documents/archive/edge53.html.

5. Richard Dawkins, *The Blind Watchmaker*, 2d ed. (New York: Norton, 1996), 13.

6. Richard Dawkins, "Sociobiology: The New Storm in a Teacup," in *Science and Beyond*, ed. Steven Rose and Lisa Appignanese (Oxford: Blackwell, 1986), 61–78, 74.

7. Dawkins, *The Blind Watchmaker*, 13.

8. Daniel Dennett, *Darwin's Dangerous Idea: Evolution and the Meanings of Life* (New York: Touchstone, 1995), 82.

9. Michael J. Behe, *Darwin's Black Box: The Biochemical Challenge to Evolution* (New York: Touchstone, 1998), ch. 2. Behe himself uses the mousetrap example in an argument against (Darwinian) reductionism, since he claims that it captures the irreducibly complexity of some biological structures.

10. Dawkins, *The Blind Watchmaker*, 13.

11. To be sure, Dennett's most famous contribution to philosophy is his emphasis on a variety of explanatory stances (the physical, design, and intentional stances), but the success of these stances is still tied to the idea of a causal mechanism. The stances are just different ways of conceiving of that mechanism. For example, Dennett claims that "intentionality" traces back to the mechanism of natural selection but that this can only be seen when we view natural selection as Mother Nature.

12. Dennett, *Darwin's Dangerous Idea*, 74–76.

13. Biology is a good example. Physiologists benefit from thinking of our gross anatomy as a system of biological machines. In fact, the president of the National Academy of Sciences, Bruce Alberts, said that at the deeper level of biochemistry, "the entire cell can be viewed as a factory that contains an elaborate network of interlocking assembly lines, each of which is composed of a set of large protein machines" ("The Cell as a Collection of Protein Machines: Preparing the Next Generation of Molecular Biologists," *Cell* 92: 291). Whether mechanisms of such complexity can really be explained in purely materialistic terms is another matter, explored in chapter 4.

14. Paul Churchland and Patricia Churchland, "Intertheoretic Reduction: A Neuroscientist's Field Guide," in *On the Contrary: Critical Essays, 1987–1997* (Cambridge, Mass.: MIT Press, 1998), 65–79. These three styles of reduction may not be exhaustive. The Churchlands's focus (and my own) is on reductions as reconceptualizations that occur as the result of new theories. It is possible that in a broader sense,

a "reduction" merely involves the unification or simplification of previously disparate phenomena that need not directly affect our conception of those phenomena. However, it is often the case that even learning that a phenomenon really falls into a more general class involves a significant reconceptualization. Thus, for Greek scientists who thought that there was a fundamental dichotomy between celestial and terrestrial mechanics, Newton's theory involves a major reconceptualization of planetary motion, since it explains it in the same way as terrestrial motion.

15. Analyses of molecular structure typically provide conservative reductions. To learn that a diamond is a crystal formed by repeating tetrahedral carbon atoms is to reconceive just what a diamond is but not to reform our basic understanding of diamonds or to eliminate them from our ontology.

16. Thanks to Del Ratzsch for this observation. It may be, as he points out, that even some conservative reductions would be very disturbing to us because we are so wedded to the idea that secondary qualities are constitutive of what a phenomenon is. Still, as time goes on, the realization that the phenomenon is essentially retained intact seems to pacify such concerns.

17. To count gravitation itself as a mechanism trivializes the issue. If the mechanists allow vast unanalyzable causal powers of this sort, they cannot claim to reject all skyhooks in favor of cranes. The whole point of a crane is that while it explains the emergence of otherwise mysterious behavior, we can also give an account of how the crane itself is built. This is precisely what Newtonian science could not do in the case of gravitation: at least at the time, gravitation was a skyhook.

18. To be sure, Newtonian physics also corrects Kepler's laws (whose description of a planet's orbit takes account of the influence of the sun but not of other planets), but it does affirm their approximate truth at least for solar systems like ours.

19. In this sense, the example demonstrates conservative reduction because, although the motion is in no sense denied, it is reconceptualized by identification of the governing force: planetary motion turns out to be "just" a gravitational phenomenon and not a unique species of motion.

20. See Michael Hoskin, "From Geometry to Physics: Astronomy Transformed," in *The Cambridge Illustrated History of Astronomy*, ed. Hoskin (Cambridge: Cambridge University Press, 1997), 116–17.

21. Newton indeed had been trained in the philosophy of Cartesian mechanism.

22. Another example of a reforming reduction is the discovery that weight, thought to be an invariant property of an object, is really a relation and varies depending on the proximity of massive bodies.

23. See the Churchlands, "Intertheoretic Reduction: A Neuroscientist's Field Guide," 70–71.

24. Indeed much larger misconceptions do not prohibit successful reference. The Greek astronomers, convinced of the immutability of the heavens, understood the comets as exhalations of the earth, confusing them with atmospheric phenomena. This misconception did not prevent them from noticing the comets or from having any true beliefs about their appearance.

25. As Dennett puts it, "Folk physics is the system of savvy expectations we all have about how middle-sized physical objects in our world react to middle-sized events" (*The Intentional Stance* [Cambridge, Mass.: MIT Press, 1987], 7–8).

26. It is no doubt true, however, that in individual cases the term *phlogiston* still sometimes refers to something real. The problem is that as a general term, it suffers from acute referential indeterminacy, since the properties it describes really belong to distinct gases.

27. See, for example, the BBC "Guide to Life, the Universe and Everything" at www.bbc.co.uk/dna/h2g2/A597152.

28. For example, Fred Dretske, Jerry Fodor, and John Searle.

29. See Dennett's "Evolution, Error, and Intentionality," in *The Intentional Stance*, 287–321. Mother Nature is natural selection conceived from the intentional stance, that is, in such a way as to license intentional ascriptions. The legitimacy of this conception will be examined in chapter 3.

30. Thanks to Del Ratzsch for pushing me to clarify this point.

31. In one sense, alchemists were eliminative reductionists because their program implies that gold and silver are not fundamental natural kinds. In another sense, they were conservative reductionists since they did not deny the existence of gold and silver.

32. Perhaps it is not correct to say that the advance of chemistry depended on the refutation of alchemy since what we would now call chemistry was blended with alchemy for some time. Nonetheless, the final discovery of the periodic table and an increased understanding of how elements behave surely refuted alchemy. If chemicals can be reduced to combinations of certain discrete elements and if (at least outside of nuclear physics) there is no means of converting one element to another, then the discovery that gold and silver are fundamentally different elements than lead and other base metals implies that the alchemists were mistaken.

33. See the Churchlands, "Intertheoretic Reduction: A Neuroscientist's Field Guide," 72.

34. For more on this, see one of the many websites on quantum nonlocality, such as www.cosmopolis.com/topics/quantum-nonlocality.html. On this site: "Our 'local realistic' view of the world assumes that phenomena are separated by time and space and that no influence can travel faster than the speed of light. Quantum nonlocality proves that these assumptions are incorrect, and that there is a principle of holistic interconnectedness operating at the quantum level which contradicts the localistic assumptions of classical, Newtonian physics."

35. The indeterminism of quantum phenomena, while certainly incompatible with Cartesian or Newtonian conceptions of mechanism, is compatible with a broader notion of mechanism, which would allow, for example, the idea of nondeterministic automata. If physically implemented, the source of the nondeterminism could be some quantum phenomenon, such as the decay of a radioactive nucleus. However, this event, while influencing the mechanism, would not be part of the mechanism itself since it still violates the requirements of determinacy and continuity. Thanks to Del Ratzsch for emphasizing this point.

36. See the Churchlands, "Intertheoretic Reduction: A Neuroscientist's Field Guide," 73.

37. Bacon, *The New Organon*, bk. 1, 54: 46. Newton, too, was obsessed with magnetism until he realized that it could not explain gravitational phenomena.

38. Hegel had supposed that ideas develop via a process of thesis, antithesis, and synthesis; and he just happened to be alive at the time of the final synthesis. It is hard to miss the similarity with E. O. Wilson's claims that an evolutionary perspective facilitates a "consilience" of all schools of thought.

39. Nagel does not mean "idealism" in the standard sense that all of reality is mind-dependent. However, what he is insightfully pointing out is that scientism identifies what can be objectively known with what can be discovered using certain mental procedures, thereby denying the real existence of knowledge that could be discovered by alternate means.

40. Nagel, *The View from Nowhere*, 9.

41. See Jerry Fodor, *The Mind Doesn't Work That Way*, ch. 5.

42. A good critique is found in Tom Bethell's, "Against Sociobiology," *First Things*, January 2001. The term *sociobiology*, tracing to E. O. Wilson, is now usually replaced with *evolutionary psychology*.

43. The issues of scientism and dogmatism in science are taken up in more depth in chapter 8.

44. I do not here assume that full agency requires a vivid awareness of why one is doing what one is doing.

45. The converse does not go without saying since intentional states might be without the self-representation presupposed by agency.

46. One sign of agency is what Del Ratzsch calls "counterflow," phenomena that apparently are not the products of the natural, unaided forces operative in the environment. See his *Nature, Design and Science* (Albany: State University of New York Press, 2001).

47. Some argue that Paley's argument should be viewed not as an argument by an analogy but as simply a comparison of the likelihood of a design hypothesis as compared with a chance hypothesis. This is a major area of controversy in ID theory. See, for example, the exchange between Robin Collins and William Dembski in *Christian Scholar's Review* 30, no. 3 (spring 2001).

48. An excellent recent discussion of British natural theology and the reason it declined is found in William Dembski's *Intelligent Design: The Bridge between Science and Theology* (Downer's Grove, IL: InterVarsity Press, 1999), ch. 3.

49. An early example is J. J. C. Smart, "Materialism," *Journal of Philosophy*, 60, no. 22 (October 24, 1963): 651–62.

50. The worry is that true versions of physicalism are likely to be vacuous because the physical is so nebulously defined that reductions are cheap or even free, a point ably defended by Tim Crane and D. H. Mellor, "There Is No Question of Physicalism," *Mind* 99, no. 394 (April 1990): 185–206.

51. Donald Davidson, "Mental Events," in *Essays on Actions and Events* (New York: Oxford University Press, 1980).

52. It is much debated in what sense supervenience is reductionist. The main problem for supervenience is that it depends on an obscure idea of noncausal determination that lacks any real explanatory power.

53. A classic early example is Hilary Putnam, "Minds and Machines," in *Minds and Machines*, ed. Alan Ross Anderson (Englewood Cliffs, N.J.: Prentice-Hall, 1964), 72–97.

54. They may also be rejecting physicalism, if they maintain that functional roles are not reducible to the physical. However, functionalism can deny physicalism without denying reductionism. A functionalist may maintain that the notion of cognitive function provides a nonphysicalist reduction of folk psychology.

55. Some, such as Grant Gillett, give powerful reasons for denying that folk psychology is any kind of causal explanatory theory. See his "Actions, Causes and Mental Ascriptions," in *Objections to Physicalism*, ed. Howard Robinson (Oxford: Clarendon Press, 1996), 81–100. If this is correct, in view of its proposed causal explanation of action, functionalism is certainly a reforming reduction of folk psychology.

56. Jerry Fodor, *The Mind Doesn't Work That Way*, 84–87.

57. Paul Churchland, "Evaluating Our Self-Conception," in *On the Contrary: Critical Essays, 1987–1997*, 25–38, 38.

58. Dawkins, *The Blind Watchmaker*, 1, 5.

59. Dennett, *Darwin's Dangerous Idea*, 214.

60. Dennett, an enthusiastic supporter of adaptionism, takes this line. See his *Darwin's Dangerous Idea*, 212–20. ID also promotes reverse engineering but takes the design to be real.

61. Elliott Sober makes the distinction between "selection of" and "selection for" in his *The Nature of Selection* (Cambridge, Mass.: MIT Press, 1984), 97–102. He says that "'selection of' pertains to the *effects* of a selection process, whereas 'selection for' describes its *causes*. To say that there is a selection for a given property means that having that property *causes* success in survival and reproduction" (100). To be fair, Sober himself does not claim that "selection for" suffices to understand the idea of biological function.

62. As we will see in chapter 3, if "selected for" is given a defensibly materialist definition, it clearly does not suffice to account for the notions of function and intentionality.

63. Fodor, *The Mind Doesn't Work That Way*, 86–87.

64. Dawkins, "Is Science Killing the Soul?" 6.

65. According to meme theory, this happens through the oral, textual, and electronic mediums, which support the copying, modification, deletion, amendment, extension, or corruption of the information transmitted.

66. Fodor, *The Mind Doesn't Work That Way*, 81.

67. See Richard Lewontin, Steven Rose, and Leon Kamin, *Not in Our Genes: Biology, Ideology, and Human Nature* (New York: Pantheon Books, 1984); and Kenneth Miller, *Finding Darwin's God: A Scientist's Search for Common Ground between God and Evolution* (New York: HarperCollins, 1999), 182–91.

68. Self-organization is especially associated with the work of Stuart Kauffman and the Santa Fe Institute. See, for example, Stuart Kauffman's *Investigations* (New York: Oxford University Press, 2000).

69. The widespread belief that such intervention is incompatible with the laws of nature is false, as I argue in chapter 8.

70. Intelligent design is a broad umbrella term, compatible with different views on the age of the earth and a variety of interpretations of the designer.

71. I am indebted to Lydia McGrew for her crystal clear analysis of this point.

72. See, for example, Dembski's *Intelligent Design*, ch. 5. A more rigorous treatment is presented in his *The Design Inference* (Cambridge: Cambridge University Press, 1998).

73. Dembski, *Intelligent Design*, 160.

74. Furthermore, a design inference is compatible with complete agnosticism about the identity and motives of the designer (one can detect a murder without knowing who the murderer is or why they did it) so that no conclusions necessarily follow about the character of the agency either.

75. They need not, of course, deny the secondary role of agents who employ artificial selection in cattle breeding and elsewhere. Likewise, a proponent of WAR could maintain the "panspermia" hypothesis, according to which life on Earth was seeded by alien intelligences.

CHAPTER TWO

~

Strong Agent Reductionism: Materialism and the Rationality of Science

> There seems no room for agency in a world of neural impulses, chemical reactions, and bone and muscle movements.[1]

> You cannot go on "explaining away" forever: you will find that you have explained explanation itself away. . . . To "see through" all things is the same as not to see.[2]

Strong Agent Reductionism

In the last chapter, I sketched a definition of our intuitive concept of agency. My emphasis was that an individual is an agent when among the individual's representations are ones that meet the following three conditions:

1. The representations exhibit intentionality: they are about something beyond themselves.
2. More specifically, the representations serve as reasons for action.
3. Still more specifically, the representations involve self-representation because they represent the individual's doing the action; they are not merely reasons for an action to be done but reasons for the individual to do it.

According to strong agent reductionism (SAR), this concept of agency is bankrupt. It merely formalizes the outmoded common-sense theory of folk psychology (FP), which is ripe for displacement by a truly scientific account of human cognition and behavior. According to SAR, what Darwin only started now needs to be finished. Darwin provided a means for eliminating agency in our natural environment: the blind automatic mechanism of natural selection. Now it is time to apply the same strategy to ourselves, replacing occult talk of human agency with the blind automatic mechanisms of neurophysiology.

It is not that the proponent of SAR holds that FP is merely incomplete or lacking in predictive accuracy so that what FP needs is correction, perhaps by way of a reforming reduction to a functional or, more specifically, computational theory of mind. Rather the claim is that FP is as radically mistaken about the facts of human cognition as both the phlogiston theory was about the facts of combustion and respiration and the caloric fluid theory was about the heating effects of friction. When a theory is that far off the mark, it is not merely that it should be superseded by a better theory, which invokes the same ontology. The ontology itself is the problem. The only proper conclusion to draw from the incoherence of the theories of phlogiston and caloric is that no such entities exist. Likewise, according to SAR, FP is so systematically misguided that the entities that it postulates—the beliefs, desires, and other intentional states of an agent—cannot survive FP's displacement by its truly scientific successor. Since these intentional states are constitutive of our intuitive conception of agency, that conception must also perish. Apparently, we must learn quite different ways of "understanding" one another.

The proponent of SAR is thus committed to some kind of eliminative reduction of agency. One approach would be to offer a diachronic account, according to which all human cognitive and behavioral capacities must trace either to our genes, via biological evolution, or to our memes (discrete memorable units), via cultural evolution. I examine this approach in chapter 5. For the present, I focus on the best-known contemporary proposal for a synchronic eliminative reduction of agency: Paul Churchland's program of eliminative materialism (EM).[3] First, I spell out the main characteristics of FP so that we have a clear picture of what EM is criticizing. Then I outline the main arguments offered in favor of EM. Next, I consider some of the main problems that can be raised for EM and show the inadequacy of Churchland's replies to his critics. Then I step back and consider fundamental problems with the very idea of SAR and, in particular, its incompatibility with the rational practice of science. If SAR is really an expression of the scientific attitude, then science is a house divided against itself, and such a house cannot stand.

Folk Psychology

The idea of a folk theory is the idea of an intuitive prescientific framework for understanding the world that all human beings have. For instance, folk physics informed carpenters and builders long before the rise of modern science, and it continues to guide our everyday physical transactions with the world when efficient practical decision making is more important than quantitative accuracy or identification of underlying causes. Folk physics may use superficially scientific terms like *friction* or *force*, yet the meaning of these terms remains vague and unanalyzed; and while folk physics is useful in giving a basic account of what has happened and in grounding rough predictions, it lacks the careful distinctions, precision, and identification of initial conditions and idealizing factors found in scientific physics.

Similarly, FP is a framework humans somehow developed for understanding themselves and others, long before the rise of serious cognitive science. It is likewise what we use in everyday transactions with other people when we have a premium on making good decisions quickly rather than a precise measurement of behavior or an analysis of its deepest causes.[4] FP trades in the terms of an agent's beliefs, desires, fears, hopes, and intentions, which can be understood as "propositional attitudes," such as the belief that *p* and the desire that *q*, and it explains and predicts human actions at the level of "common sense." A good source of FP explanation and prediction is literature, with mystery fiction an obvious example. Thus, we might read the following and then provide an FP explanation:

> Due to mismanagement of his widget factory, Hagarth-Smythe was broke yet still addicted to the comforts of aristocracy. He earnestly desired to remain in the lifestyle to which he had become accustomed and believed that murdering Colonel Braithwaite-Stewart would ensure that Aunt Agatha's family fortune would come to him. Finding the opportunity to lace Braithwaite-Stewart's single malt with arsenic, Hagarth-Smythe did the dirty deed.

In light of Hagarth-Smythe's desire and beliefs about how to satisfy that desire, his murderous action is a rational one.[5] Not only does FP explain his action, it can also predict others. Suppose Inspector Thatchwood concludes that Hagarth-Smythe is the culprit but also realizes that Cynthia Holmbury-McAlpine must know this as well because she saw Hagarth-Smythe leaving the woodshed where the arsenic is kept, bottle in hand. Furthermore, the inspector sees that Hagarth-Smythe suspects that Holmbury-McAlpine knows because she asked him what he had been doing in the woodshed the day before the murder. In view of Hagarth-Smythe's evident and ruthless desire to remain wealthy, Thatchwood predicts that Holmbury-McAlpine is in danger of being Hagarth-Smythe's next victim and rushes over to the family estate hoping to avert catastrophe. Not only in fiction but also in real life criminal investigation, FP may be useful enough to save someone's life.

The analogy between folk physics and FP makes it seem obvious to the proponent of EM that FP, like folk physics, is something that has a strong chance of being displaced, at least as an accurate scientific account of the world. Naturally, neither FP nor folk physics may be practically dispensable in everyday life. No doubt we retain folk physics for ordinary life not only because few of us are trained scientists but also because it works well enough for the rough-and-ready approximations we need for typical household tasks, driving automobiles, and participating in team sports. But the evidence is that folk physics is only a useful fiction that happens to approximate the truth for the typical environments that people find themselves in. The fact is that folk physics is seriously in error about many things:

- it believes that weight is intrinsic, when actually it is relational;
- it has a naïve idea of the solidity of an object, ignoring the fact that most of the object is empty space—say, for example, a table;
- it retains the common-sense idea of centrifugal force (whether it uses the term or not), even though the notion is fictional and redundant;[6]
- indeed, it often talks of force as though it were a homogenous thing, like vanilla ice cream, something from which more specific forces can be obtained by a variety of "flavorings."

Thus, it may be argued that much of what folk physics trades in are convenient fictions, concepts that enable us to manipulate our environment effectively but whose specific terms do not refer to distinct, well-defined entities recognized by science.

On the strength of this analogy, proponents of EM are prepared to argue that FP is an intuitive theory constructed from convenient fictions about the motivation and behavior of human beings. These fictions work very well within a limited domain of human behavior, which assumes normal biological and psychological development, a certain level of maturity, and the absence of damage or disease in the brain and nervous system. However, if a deeper theory of human motivation is developed, one that takes seriously the actual patterns of neural activation in the brain, it may be possible to give a general uniform account of human behavior that works in the intuitively rational ("normal") cases and in the irrational ("subnormal" or "abnormal") cases. It seems to proponents of EM that a deeper theory of this kind will not need to trade in the beliefs, desires, and other "reasons" of FP, which gain credibility only because they work well enough in the normal cases. In fact, the proponent of EM thinks the whole idea of "reasons" for action that is enshrined in FP misleads people into postulating fictitious propositional attitudes, such as the belief that *p* and the desire that *q*. By analogy with a public argument in which propositions are laid out as reasons to believe some conclusion, it is supposed that our mind contains propositional attitudes that combine in a practical syllogism to produce action. According to EM, the idea of propositional attitudes is misguided because the alleged propositional content fails to model what actually occurs at the level of neural-activation patterns.[7] Here, information is apparently distributed among various nodes in nonpropositional form,[8] and one can gain a much more precise picture of expected patterns of behavior by attending to the details of the prior activation patterns and their method of transformation. Notice, therefore, that EM is certainly not offering a vacuum in place of propositional attitudes. EM does not deny that important structured information processing is going on; it simply maintains that the propositional attitudes of FP do not provide a scientifically promising model of how this processing is accomplished.

At the same time, defenders of FP would call attention to what appear to be important disanalogies between FP and folk physics. Foremost of these is that FP is integral to our self-conception as rational beings (and agents), whereas folk physics is not. Learning that our common-sense conception of the physical world is radically in error, in need of serious reform, and sometimes in need of outright elimination is disconcerting; but it is not a direct threat to our identity, autonomy, and dignity. However, regardless of its incompleteness and lack of precision as a general theory of human psychology, FP does capture the idea that, at least often (in normal circumstances), human beings act for identifiable reasons. Reasons, however, seem intimately tied to propositions. Regardless of whether the context is psychological or not, when we ask for the reason for something, we expect an answer in the form of a proposition, or a completed thought,[9] as when we ask for the reason that the bridge collapsed and expect an answer such as "The hawsers were fatigued." The distinctive character of psychological cases is that it seems appropriate to say that an agent "has" reasons: the reasons do not merely exist but serve as an internal motivation for the agent's further thought and action. But unless there is a curious equivocation going on when we use the word *reason*, reasons will not cease to imply propositional content simply because the reasons are internal to an agent. If this is right, then when the proponents of EM argue against propositional attitudes, they seem to imply that agents do not act for reasons. But if that is the case, rejection of FP does threaten our self-conception as rational beings in a way that the rejection of folk physics does not.

Other points are in favor of FP, some of which are brought out in due course, but the apparent linkage of FP to the idea of an agent's rationality is surely the most important reason that the claims of EM are vigorously contested. However, before we state our critique of EM, we should examine the strongest reasons that have been advanced in its favor.

The Case for Eliminative Materialism

One of the critical assumptions of EM is that human self-understanding is not merely a skill or deliverance of introspection; it is a collectively internalized empirical theory. Thus, although it is not scientific in any strict sense, FP does have significant empirical content, and so it is not overtly absurd to suggest that FP could be refuted on the basis of observation. The idea that FP is an empirical theory traces back to a classic paper by Wilfrid Sellars.[10] Using a mythical account of human development, Sellars argued that folk psychology arose when human beings started to use language not merely to describe external reality but as a means of characterizing their mental states. By employing such "propositional attitudes," it became possible to understand how people's thinking provided reasons for their action. The same kind of proposition that could be used in public reasoning (e.g., in a spoken or written argument) could now be used to capture an individual's private reasons for action. Not only does this model allow third-person explanation of action, it also facilitates first-person accounts of action that are consistent with a third-person perspective. Since FP enables human beings to understand one another and has withstood the test of many centuries of use, Sellars was happy to conclude that FP is basically correct.

Churchland accepts Sellars's argument that FP is an empirical theory but argues instead that FP is false. Four arguments are offered for rejecting FP.

First, FP is woefully incomplete as a psychological theory: "FP fails utterly to explain a considerable variety of psychological phenomena: mental illness, sleep, creativity, memory, intelligence differences, and the many forms of learning, to cite just a few."[11] Indeed, the charge has considerable force because FP focuses on the reasons of an agent whose cognitive system is in good working order. FP says nothing about how this reasoning is implemented, what happens when the implementation goes wrong, or about various nonrational psychological activities of human beings. While defenders of FP may see these as issues that FP is not and should not even be trying to address, Churchland demands a unified account of all human psychological phenomena and therefore sees them as evidence of FP's incompleteness and proneness to displacement. In particular, Churchland thinks that FP's failure to explain learning is particularly telling because the ability to manipulate propositional attitudes, presupposed by FP, is itself something learned and thus cannot be one of the fundamental psychological capacities we are born with.[12] In one way, this point is controversial, as it is not clear that the ability to formulate propositional attitudes—for example, desires—is a prerequisite for their correct attribution.[13] However, Churchland does call attention to the impressive results obtained from training neural networks and suggests that they may capture a more fundamental level of information processing.[14]

Second, FP is a stagnant theory, facilitating no significant new insights for 2,500 years. Churchland's idea is that, for example, the practical syllogism of FP is essentially unchanged since the time of Plato and Aristotle, who were both aware of it. Defenders of FP would suggest that basic reasoning is not the kind of thing that does change dramatically—for example, aside from metatheoretic results, such as Gödel's theorem, basic arithmetical reasoning shows only modest innovations. Churchland, by contrast, would see this example as a sign of intellectual stagnation, of what Imre Lakatos called a "degenerating research program."[15] Not only that, although FP is a theory whose main focus is human beings, it is really the last vestige of a declining theory that once had greater scope; indeed, this is the very intuition driving SAR: "FP has been in steady retreat . . . as intentional explanations have been withdrawn from yet one domain after another—from the heavenly bodies, from the wind and the sea, from a plethora of minor gods and spirits, from the visitation of disease."[16] Agency, the materialist thinks, has been expunged everywhere else in the cosmos, so it is reasonable on inductive grounds to conclude that we are next.

Third, Churchland argues that FP "shows no sign of being smoothly integrable with the emerging synthesis of the several physical, chemical, biological, physiological and neurocomputational sciences."[17] By contrast, neuroscience has strong connections to this synthesis. Actually, this point is probably Churchland's least persuasive because, as I argue in more detail later in this chapter, FP is implicated in the very method of science itself and so could claim support from any science employing that method.[18] If all scientists think of themselves as employing propositional attitudes (and must think of themselves as doing so), the case can be made that FP is sure to be included in any synthesis of ideas that have contributed to scientific success, regardless of whether it is the subject matter of many successful scientific theories.

Churchland puts most weight behind a fourth objection to FP. If FP is an empirical theory, it finds no confirmation in our best neuroscientific models of the brain. According to FP, an individual's thinking is propositional in character, and the decision making that results involves the inferences of practical reasoning. Agents desire certain goals, believe that certain actions tend toward those goals, and thus decide to perform those actions (other things being equal). However, when neuroscientists investigate the workings of the brain, the most fruitful model is the one provided by neural networks that employ parallel distributed processing. According to this model, representations do not have the character of identifiable propositions. Rather, they are patterns of neuronal activation that may be interpreted as "activation vectors." Learning depends not on storing sentencelike entities[19] but on the development of prototype vectors where concepts are represented by modeling the distance of various items from prototypical exemplars. Furthermore, when the brain's activity leads to action, nothing corresponds to any sort of inference. Rather, a vector-to-vector transformation is happening between many layers of neurons.

> In this way, a sensory activation pattern can undergo many principled transformations before it finally finds itself, profoundly transformed by the many intervening synaptic encounters, reincarnated as a vector of activations in a population of *motor* neurons, neurons whose immediate effect is to direct the symphony of muscles that produce the coherent bodily behavior appropriate to the original input vector.[20]

In this model, beliefs, desires, and other intentional states seem to disappear. Still, even if such intentional states do not exist, Churchland has acknowledged that the theory FP does exist. So he must give some account of the origin of FP. His suggestion is that "FP, like any other theory, is a family of learned vectorial prototypes that sustain recognition of current reality, anticipation of future reality, and manipulation of ongoing reality." Of course, in the case of FP, the learned vectorial prototypes model cognition in terms of propositional attitudes. But since they are only models, "there is no reason why [FP] must be correct in so representing our cognition, nor in representing itself in particular."[21]

In short, Churchland argues that FP is an incomplete, stagnant vestige of a prescientific worldview that is incompatible with our best theories of brain function. Due to that failure of fit, there is little hope for a conservative or even a reforming reduction of FP. FP, along with the intentional states it postulates, simply has to go.[22]

The Robustness of Folk Psychology

Churchland's claims about FP have provoked many critical responses. Putnam and Dennett have argued that FP is not a theory but a craft (or a skill or practice) indispensable to human interaction so that revision of our theories about FP would not provide a practical alternative to the craft.[23] Greenwood and others have challenged the idea that FP is stagnant because in fact contemporary social psychology applies, deepens, and extends FP's explanatory categories: "It advances intentional explanations of human actions undreamt of by the medievals."[24] Many have pointed out that the incompleteness of FP simply means it has a limited domain and can be supplemented with other theories. Some have claimed that FP cannot be falsified by neuroscience because it abstracts from an identification of the specific causes of an action so that it is consistent with neuroscience's providing a lower-level account of the phenomena. Heil suggests that instead of viewing activation vectors as models of psychological representation, "were networks taken . . . as *implementations* of psychological theories—models of their neuronal realizations, for instance—then folk theorists might be perfectly happy to concede that causal principles operating at the underlying level could be connectionist in character."[25]

These points are all inconclusive. If a craft is a system of habits of mind, it still needs to be asked whether they are good habits. The ubiquitous use of induction fails to justify the practice. Even if a craft is not a theory, it is still worth asking if the codification of a craft as a theory is true. If it is not, then Churchland will urge that refusal to modernize the craft is a mere Luddite resistance to the advances of science. Surgeons were once in the habit of not bothering to sterilize their hands. If the practice is codified as the theory that sterilization is not important to surgery, the falsity of the theory also tells us that the practice had to end. That FP can be extended to social psychology is good news, but Ptolemy's remarkable theory could also be significantly extended by the addition of more epicycles. We still have good reason to think that epicycles do not exist. Powerful useful theories can be false and postulate nonexistent objects. That FP is incomplete as a psychological theory is compatible with supplementation but also with subsumption under (or displacement by) a more general unifying theory.[26] Finally, FP might be compatible with a connectionist theory of mind in the way Heil suggests, but not only does that sound like an empirical question, their mere compatibility does not imply any confirmation of FP from neuroscience—and that such confirmation does not exist is precisely Churchland's point.[27]

However, although the eliminativist seems to have effective responses in these cases, other objections to EM have greater power. In this section, I present four main objections, then I devote another section to the fifth and most important objection: that EM (and, more generally, SAR) is incompatible with the rational practice of science.

Objection 1: The Abstraction Problem

FP seems to uncover real patterns in human behavior that are invisible at the level of physical sciences, such as neurophysiology. This is a case where the neurophysiological crane loses contact with an important *explanandum*.

Dennett sees the value of FP as authorizing a distinct explanatory stance, the "intentional stance." He argues, and Churchland would agree, that we currently have no practical alternative for understanding one another since the theory is embedded in our expectations in a way that makes it vastly more efficient and accessible than any neuroscientific theory. That position, however, is consistent with the intentional stance's being a useful fiction, which at one time seems to have been Dennett's view. Since then, Dennett has become a "quasi realist." While he does not accept a traditional understanding of intentionality (see chapter 3), he argues that adopting the intentional stance reveals real patterns that are invisible from a purely physical stance. Consequently, even if FP stands in need of revision, Churchland's case for elimination fails because FP gets at realities that neurophysiology has no clear means of capturing, at least if it is confined to the physical stance.[28]

To make his case, Dennett pits ordinary humans using FP against Laplacean superphysicists[29] from Mars, both attempting to account for the behavior of stockbrokers on Wall Street. To apply this example specifically to Churchland, let us suppose that these physicists can trace all the signals entering a stockbroker's brain; can compute billions of the brain's vector-to-vector transformations, even quicker than the brain itself; and can consequently "predict the exact motions of his fingers as he dials the phone and the exact vibrations of his vocal cords as he intones the order."[30] In terms of providing precise, accurate predictions of specific bodily movements, the Martians leave the humans in the dust. However, the humans see something real that the Martians do not, namely, the irrelevance of many of the specifics to understanding the stockbroker's action. Suppose that what the stockbroker did was sell 100 shares of Company XYZ. Not only could the specific finger motions and vocal-cord vibrations have been different, but the stockbroker could have done the same thing by using a different phone, by making an online transaction, or by asking someone else or even a company to do it. As Dennett says, "If the Martians do not see that indefinitely many *different* patterns of finger motions and vocal cord vibrations—even the motions of indefinitely many different individuals—could have been substituted for the particulars without perturbing the subsequent operations of the market, then they have failed to see a real pattern in the world they are observing."[31]

If Churchland's activation vectors prove as successful in real life as they are in the hands of the Martians, they will be wonderful cranes for explaining the etiology of specific human movements. But they seem to lack the abstraction necessary to capture the true *explananda* of FP, which aims to explain a general result (e.g., the action of selling shares) and is indifferent to how this is done. At this point, it would surely be implausible for Churchland to claim that such actions are not legitimate scientific categories because they are too abstract to be captured by neuronal activation vectors. Actions do not merely figure in FP; they are also presupposed by sciences such as economics. Economics needs the concept of action for a reason independent of FP:[32] it has the right level of abstraction for understanding *economic transactions*, which as we all know can be accomplished in many different ways. To the economist developing a model of market behavior, all that matters is whether the stockbroker bought or sold, not the sequence of movements that implemented that transaction. To deny the legitimacy of actions would deprive economics of its *explananda* and thus its status as science.[33]

Once actions are admitted into our ontology, however, it is hard to dispense with intentional states. Consider an analogy between actions and sunburn. Just as sunburn points to a certain etiology, action classifications point, if not to any particular beliefs and desires, at least to the intention of some agent. Each particular case of sunburn can be individuated by its intrinsic properties quite apart from its cause; but each of these cases is also distinct, so it is hard to see what else they can have in common besides their cause. Likewise, buying or selling shares is understood as something an agent planned to do. Each action can be individuated intrinsically, but because of the open-ended variety of cases, it is hard to see how the class of actions can be reconstructed without appeal to a common intentional cause. Certainly, it is not clear how this can be done by appeal to activation vectors. For one to explain the variety of possible movements that can implement an action, one would have to appeal to a variety of quite different activation vectors.[34] If this can be done, it would require a high degree of abstraction capturing a whole class of such vectors. Yet because the activation vectors are physically so different from one another, it is not obvious that the class could be individuated without appeal to the fact that all its members implement the same intention of the agent. The more abstract the specification of the activation vectors, the more it seems likely that we will invoke precisely those intentional qualities EM aims to dispense with.[35]

Among proponents of FP, opinion is divided on whether intentional states provide a causal explanation of actions.[36] But even if they do not, the abstraction problem provides evidence that the postulation of intentional states is needed to identify actions, and this position is incompatible with the program of SAR in general and EM in particular.

Objection 2: The Subjectivity Problem

That our mental lives have a subjective dimension is an undeniable fact that is invisible in the objective third-person accounts of neurophysiology.

Searle has pointed out an inevitable limit to the standard reductive strategy practiced by modern science.[37] While he leaves open the possibility of "a new—and at present unimaginable—conception of reduction,"[38] his argument has force against EM because EM is wedded to the standard strategy. The argument depends on understanding the limitations of the distinction between appearance and reality.

Searle notes that science has made considerable progress by dividing the world of experience into objective, primary qualities, such as motion and extension; and subjective, secondary qualities (appearances), which arise from the interaction of primary qualities with our senses. For an objective view of heat, one's personal feeling of heat has to be factored out, and the objective cause of that sensation (molecular motion) has to be identified. This dichotomy works because it is possible to make a clear distinction between subjective appearance and external reality. However, when we turn to an analysis of our own mind, this distinction breaks down. When I consider the feeling of pain (the appearance), it is no use pointing out that certain neurological events are occurring in my thalamus. Although the events tell us about the primary qualities that produce the feeling of pain, they cannot be what that feeling actually is. The feeling is a secondary quality, and the whole point of introducing primary qualities was to distinguish them from secondary qualities. In other words, the neurological events are third-person objective events, but the feeling of pain is intrinsically first personal. It does not work to say that the neurological events are the reality, the feeling a mere appearance, because in this case, "*the appearance is the reality*."[39]

Surprisingly, Churchland has not recanted. According to Churchland, Searle's argument is either a non sequitur or question begging.[40] Churchland concedes the difference between first- and third-person perspectives, but he maintains that the difference is epistemic, not ontological. An individual's sensations and thoughts are known to him or her in one way, from a first-person perspective; but the corresponding neurological events are known in a different way, from a third-person perspective, say, to a neurophysiologist. It is nonetheless a fallacy to argue that what we misleadingly call the sensations and thoughts are not really just the neurological events. This position, according to Churchland, would be as bad as arguing in the following way.

1. The temperature of an object is known to John by simple feeling.
2. The mean molecular kinetic energy of an object is not known to John by simple feeling.

3. Therefore, since they have divergent properties, temperature cannot be identical with mean molecular kinetic energy.[41]

Searle can surely respond that Churchland has missed the point, especially as he prefaced his remarks by insisting that "the point of the argument is ontological and not epistemic."[42] Searle's argument is not that different properties are discernible from the first- and third-person perspectives, it is that the character of the discernment itself is different. It is not what is sensed or thought about but the nature of the sensing and thinking that is decisive. In Nagel's terms, first-person sensations and thoughts have an internal what-it-is-like quality that cannot be discerned from a third-person perspective. That John knows about temperature in a distinctive way from the way it is understood by mechanics does not show that temperature is anything other than what mechanics says. But it is a reason to think that first-person ways of knowing cannot readily be assimilated to third-person perspectives. The difference in ways of knowing seems ontological.

In fact, Churchland concedes that Searle intends to state an ontological, not an epistemic, difference between first- and third-person perspectives.[43] But Churchland claims that if Searle does so, he is just begging the question:

> Whether or not the subjective mental features one discriminates in introspection are identical with some objective features of one's brain that might eventually be discriminated in some objective fashion is exactly what is at issue. What will determine the answer to the question is not whether our subjective properties intuitively *seem* to be different from neural properties, but whether cognitive neuroscience eventually succeeds in discovering suitable systematic neural analogues for all of the intrinsic and causal properties of mental states.[44]

It is clear that Churchland does not see Searle's point. First, talk of "subjective mental features one discriminates" changes the subject by employing a third-person description of the contents of subjective experience. It is not what is discriminated but the subjective nature of the discrimination that is problematic for EM. Furthermore, it does not help for Churchland to suggest that subjective properties may only seem different from neural properties. This position covertly retreats to the epistemic understanding of Searle's argument. The problem for EM is not that subjective properties seem different from neurological ones but that seemings are tied to a particular point of view. It gets us nowhere to suggest that we only seem to have distinctive points of view since this seeming itself presupposes such points of view.

When appearance is the object of study, saying that the appearance only appears different from objective reality only provides more evidence for the existence of appearances and for their irreducibility to objective accounts of reality. It is useless to counter by claiming that the appearances are only an illusion since illusions are themselves appearances. As Descartes might say, "Deceive me with any appearance you may, you will never deceive me into thinking there are no appearances or that there is no ultimate distinction between appearance and objective reality."

What this argument shows, I believe, is precisely what Nagel has emphasized.[45] Reality is not synonymous with the "objective" view characteristic of science.[46] That view was designed to enable us to study the characteristics of objects, such as temperature, independent of their impact on human subjects. It was never intended to deny that human subjects exist. Indeed, it is only subjects who are able to engage in science and adopt the third-person point of view. Only something that already has such a thing as a first-person point of view can so much as adopt a third-person perspective. The possibility of distinguishing appearance from objective reality presupposes that appearances are also dimensions of reality.[47] The objectivity of science is parasitic on a prior recognition of subjectivity, and thus it is implicitly self-refuting to use science to eliminate the subjective point of view. Subjectivity is a part of reality that does not reduce to the objective view of science but that is nonetheless indispensable to and presupposed by science. Churchland's whole mistake is that he thinks that the viewpoint of scientific theory can be used to establish conclusions independently of the presuppositions of adopting that viewpoint.

The reality of subjectivity is evidence of the correlative notion of agency, since it is agents who have particular points of view and, thus, beliefs and desires directed toward their own actions. The irreducibility of subjectivity is thus another argument against SAR in general and EM in particular.

Objection 3: The Ontological Robustness Problem

The falsity of FP as a scientific theory would not justify the elimination of intentional states from our ontology.

Churchland subscribes to a typical empiricist position on the semantics of theoretical terms. According to this view, the meaning of such terms is specified by the causal roles they are ascribed by the various "laws/principles/generalizations in which they figure."[48] Consequently, theoretical entities exist only if the specified causal roles are actually played out. Perhaps some minor revisions of our scientific theories are consistent with the retention of the entities they postulate, but if there is reason to think that nothing plays anything like the causal role ascribed, we should conclude that no such entities exist. According to Churchland, intentional states are theoretical entities postulated by FP, but there is no convincing empirical evidence that anything plays the causal role ascribed to intentional states in FP's accounts of practical reasoning and action explanations. Thus, Churchland concludes, no such intentional states exist.

One might object that even if Churchland's premises were correct, what would follow is only that we should deny the existence of intentional states as described by FP. It is consistent with this position that intentional states exist but are different than we supposed. They might, for example, live on via a reforming reduction to a functionalist theory. However, this angle does not seem to be Churchland's point. His opposition to FP is based precisely on the fact that it is a species of the intentional explanation that has been in "steady retreat,"[49] and so the problem with FP is that it postulates states with intentional content. If nothing plays this role, then Churchland's point has to be that it is the very idea of intentional states that has to go.

But Churchland's argument is not correct. Churchland's account of theoretical entities is motivated by examples where the entities are purely theoretical so that there is no reason independent of the theory and its success to think such entities exist. This is a reasonable model for postulates such as epicycles, equant points, and crystal spheres. When the classical theories of astronomy were superseded by simpler theories that did not need to invoke such arcane abstractions, observers had good reason to think that these abstractions did not exist. No work that needed to be done required their existence. But entities can be theoretical without being purely theoretical and can thus survive a major reformulation of theory. For example, according to folk physics, weight is an intrinsic property of bodies: it is just "common sense" that if we do not add anything to, or subtract anything from, a boulder, it must weigh the same anywhere in the universe. Thus folk physics assigns weight the causal role of a constant and makes predictions accordingly. Unfortunately, Newtonian physics shows that the "laws/principles/generalizations" folk physics provides for weight ascribe causal roles that frequently do not obtain. The fact of gravitation means that weight is a relation between an object and other massive bodies (such as the nearest planet), and so, as every student knows, the same boulder is heavier on Jupiter than it is on Earth. If we assume that boulders always weigh the same as they do on Earth, our predictions of their behavior on Jupiter will fail. Yet, with cavalier disregard for empiricist philosophy of science, no one concludes that weights do not exist. The reason is obvious. We have all sorts of reasons independent of any theory, even folk theory, for thinking that weight is something real. Weights have clear detectable effects and are remarkably stable in the particular environment we inhabit, a fact that fair trade, engineering, and all sorts of scientific experimentation depend on. That weight turns out to be relational and not intrinsic merely calls for a reform of our theoretical conception of weight, not for an elimination of weight from our ontology.[50]

It is thus not enough for Churchland to claim that intentional states are entities postulated by a false theory. He also needs the assumption that intentional states are more like equant points than weights. To refute this assumption, I need to show that we have evidence for the existence of intentional states that are independent of FP. To do so might seem like a tall order since FP is such a pervasive and internalized means of mutual understanding. In fact, it is possible to identify at least three distinct kinds of independent evidence for intentional states.

The first kind of evidence is empirical. As Greenwood notes, developmental psychologists[51] have found that very young children begin with a sharp distinction between causation, which is attributed to concrete items, and their own mental life. Only later and through experience do children learn that mental states involve causes and effects of observable events. These mental states include intentional states such as the desire for breakfast, the fear of dark, and the love of play. Four-year-old children learn to use language to ascribe such states to themselves and others, but it is surely false that they do this on the basis of FP, conceived as a developed prescientific causal–explanatory theory. These children are not postulating mental states on the basis of acquiring FP but ascribing causal roles to mental states whose existence they already acknowledged as a means of making sense of their own experience of interacting with others. It is not merely that they could not state FP as a theory (which by itself proves nothing because it could be an implicit theory, as Churchland would urge); rather, it is that such a theory was not their reason for attributing causal roles to mental states: "And if this is so, it is surely possible for adults and psychological scientists to ascribe psychological states to self and others meaningfully and to have evidence for them (via talk), even if their causal explanatory theories of discourse and action turn out to be inaccurate or inadequate."[52]

Of course it may be said that even the perceptions of children are "theory laden," but the theory in question seems to be that of common sense. While the deliverances of common sense are often radically revised by science—for example, objects do not have colors or weights in the way common sense supposes—the ontology is typically robust. And as Searle emphasizes, the fact that common-sense objects do not figure in refined scientific theories is a red herring. Such common-sense entities as cocktail parties and split-level ranch houses go completely unmentioned in physical theories, but that does not count against their existence. "In a word, beliefs and split-level ranch houses are totally unlike phlogiston because their ontology is not dependent on the truth of a special theory, and their irreducibility to a more fundamental science is irrelevant to their existence."[53]

The second consideration is logical. Churchland's understanding of FP as a theory depends on a crucial claim of Wilfrid Sellars. Sellars's reason for claiming that FP is a theory depends on the assumption that language use preceded the development of FP. The decisive moment in the development of FP was when humans started using language as a model not merely for external reality but also for their own mental lives. According to Sellars and Churchland, FP is a theory because it is an internalized linguistic model. It was during the internalization that intentional states were postulated. This proposal, however, has a clear logical difficulty that can be seen in the dilemma that arises when we ask the following question: In the period before humans internalized FP, did they understand each other's linguistic behavior (e.g., speech)? If they did not, then they would not have been able to see that language provided a useful model for our mental lives. If I do not understand "It is raining," on what basis can I propose that someone else believes or fears that it is raining? Surely one must grasp a linguistic content (a proposition) before one can coherently propose that our mental lives involve our having a relation toward such contents (propositional attitudes). However, if humans did understand each other's linguistic behavior, then they had a pretheoretic commitment (even if an unconscious one) to something like understanding that *p*, a propositional attitude. It is therefore more plausible that the theory of propositional attitudes gave humans a means of understanding what their preexisting understanding was. If that is the case, then intentional phenomena such as understanding preceded FP and are presupposed by the ability to formulate it. But then the falsity of FP cannot show that such intentional states do not exist.

The third source of independent evidence for intentional states calls attention to their theoretical value beyond FP. One of the worst features of the Sellars–Churchland thesis about the origin of FP is that it is flatly inconsistent with one of linguistic philosophy's greatest triumphs, the analysis of the meaning of utterances in terms of the intentions of their speakers.[54] This program cannot even get started without the assumption that intentional states are more fundamental than linguistic behavior.[55] Furthermore, other sciences, such as social psychology, criminology, and cultural anthropology, have flourished by retaining and refining FP's commitment to intentional states. It is hard to see what workable replacement neuroscience could offer these sciences.

Objection 4: The Incoherence Problem

The denial of the existence of intentional states is incoherent.

A common reaction to EM is to conclude that it cannot be true because it is self-refuting. Common-sense formulations of the objection call attention to the pragmatic paradox that proponents of EM appear to believe that there are no beliefs and that they want others to believe this, too. Lynne Rudder Baker has developed the objection into a rigorous philosophical response:

> To deny the common-sense conception of the mental is to abandon all our familiar resources for making sense of any claim, including the denial of the common-sense conception. . . . In the absence of a replacement, it is literally inconceivable that the common-sense conception of the mental is false. . . . If the thesis denying the common-sense conception is true, then the concepts of rational acceptability, of assertion, of cognitive error, even of truth and falsity are called into question.[56]

Notice that Baker's objection to EM cuts much deeper than the pragmatic paradoxes noted by common sense. It is not merely that we have difficulty describing our commitments without the concept of belief and so have no obvious way to express the idea that there might not be beliefs. It is that there is no way to make sense of our rejecting FP and accepting a neuroscientific replacement that does not retain the very concepts of rejection and acceptance (intentional activities) that we claim to have dispensed with.[57] For that matter, there apparently is no such thing as making sense of anything, since making sense of something is yet another intentional activity. Nor apparently can we keep a grasp on assertion, since there is nothing specifiable to assert (such as a meaning or a content).[58] But truth is parasitic on meaning. Meaningless items can be neither true nor false. So if there is no such thing as finding or asserting a coherent meaning, there is no such thing as achieving a state of mind or making a statement that can be either true or false.[59] Thus if EM is true, it would appear that no one could know it or expect to convince anyone else of that fact. Is the proponent of EM left only with Wittgenstein's inarticulate cry?[60] It would seem that Churchland has replaced the skyhook of intentional states with the crane of neural-activation vectors only to render inscrutable the whole world of human cognition and communication. The crane has become disconnected from the *explanandum* it was supposed to lift.

Surprisingly, Churchland has not recanted. He does not deny that EM currently appears incoherent, but he does think this is simply a consequence of how pervasive FP is in our current conception of the world. Churchland has offered three arguments to neutralize the objection that EM is self-refuting. First, he points out that all reductio arguments "presuppose what they deny," since they establish that a proposition is false by showing that if we assume it to be true, we can derive a contradiction. Thus, the fact that FP is assumed in the argument against FP is perfectly legitimate, and to block this kind of argument "would signal a blanket refutation of all formal reductios."[61] A second and related argument charges that those who claim we cannot move beyond FP without incoherence are employing principles that would justify retaining any cognitive theory, no matter how bad, so long as it happened to be in pervasive current use. Supposing there had been a more primitive theory of "gruntal attitudes," prior to postulation of propositional attitudes by FP, the same reasoning could have been used to block the transition to FP.[62] A final argument makes a Kuhnian point about what happens in a paradigm shift. Abandoning FP seems incoherent now because we lack the alternate perspective of a successor theory. "We need only construct it, and move in. We can then express criticisms of FP that are entirely free of internal conflicts."[63]

Despite an air of plausibility, each one of Churchland's arguments fails. First, in his account of reductio reasoning, Churchland omits a crucial fact. In a correct use of reductio ad absurdum, the working assumption that is shown to be logically absurd is discharged, which means that the conclusion, namely, that the assumption is false, does not depend on that assumption but only on our original, given premises. This fact is shown visually in the Gentzen system of natural deduction, by presenting the conclusion of a reductio as falling within the scope of the main derivation and not merely the scope of the subderivation under the assumption to be refuted (see figure 2.1).[64]

Line #		
1 to k	premises of main argument	
n.	P	// assumption to refute
o.	Q	// contradictory
p.	~Q	// pair
q.	~P	// n-p, by *reductio ad absurdum*

Figure 2.1. Reductio ad absurdum in the Gentzen system of natural deduction.

However, not all reasoning that follows the same apparent form can be accepted as a correct use of reductio. For the most extreme case, consider a reductio argument to the conclusion that reductio reasoning is invalid. Suppose I assume that reductio reasoning is valid and that I derive some contradiction. According to the formal rule, I can now infer that reductio reasoning is invalid. But I can do so only if I employ the very reductio rule I claim to be rejecting. At just the point I claim to conclude that reductio is invalid, I am relying on the validity of reductio. The problem, evidently, is that although my argument has the right form for a reductio, it cannot be a correct use of the rule, because in drawing the conclusion, I have not really discharged the assumption to be refuted. Among my reasons for claiming that reductio was invalid is reductio itself, so if my conclusion is correct, I am not entitled to draw it (see figure 2.2).[65]

This point has everything to do with Baker's real objection to Churchland. Baker never claimed that it is always wrong to presuppose what you go on to deny. Her point was that it makes no sense to continue to presuppose it at the very point at which you claim to deny it. The problem for Churchland is that any reductio of FP will ask us to accept the conclusion that FP is false but that very notion of acceptance is one that still presupposes FP. Thus, even though an argument with the form of a reductio may be leveled against FP, it cannot be a valid use of reductio, because FP is not discharged in drawing the conclusion that FP is false. Indeed Baker gives an example of this very point:

Line #		
1 to k	premises of main argument	
r.	*Reductio* is valid	// assumption to refute
s.	Q	// contradictory
t.	~Q	// pair
u.	*Reductio* is not valid.	// r-t, by *reductio ad absurdum*

Figure 2.2. An absurd use of Reductio ad absurdum.

> Suppose that the skeptic tried a kind of *reductio ad absurdum* of the commonsense conception by using, say, the notion of rational acceptability in order to show that that notion has insurmountable internal problems; from this, he concludes, so much the worse for rational acceptance. It would remain unclear how any such argument could have a claim on us. We obviously could not rationally accept it.[66]

The only way that Churchland can avoid this argument is by actually constructing an alternative to the notion of acceptance and by "convincing" us of its legitimacy.[67]

Churchland's second response also does not work. Baker's argument is not that we cannot make a transition from FP because it is currently in pervasive use. It is that we cannot claim that such a transition is possible without giving an account of the new paradigm that avoids commitment to the concepts of FP. Churchland's mythical account of the transition from gruntal to propositional attitudes does not convince either, since his description assumes that the gruntal folk already possess intentional states. He imagines that "some forward-looking group sets about to develop a new and better conception" and that they do this by "contemplating the shortcomings of their older conception."[68] But "setting about," "conception," and "contemplating" are intentional notions, so if the gruntal folk can do these, then FP is already true of them.[69] The problem is really the same as the general problem for EM: it needs actually to show how folk-psychological concepts can be dispensed with.

In Churchland's third response, he suggests that as soon as we had a replacement for FP, it would provide a standpoint from which we could see the problems of FP without incoherence. He thinks this point is analogous to major advances in science where a new idea seemed utterly irreconcilable with an old one until we made the transition; then the scales fell from our eyes and we began to see what was wrong with the old idea. The theory of aether turned out to be utterly irreconcilable with the properties of electromagnetism. The prevailing perspective of mechanism effected no conceivable transition to Maxwell's electromagnetism theory. However, once the transition to Maxwell's theory had been made, it was possible to see the untenability of the aether theory and the limitations of mechanism. Might not FP go the same way as aether theory?

In reply, we may begin by noting the important differences between the two cases. However wrenching the transition to Maxwell's theory, there was an underlying continuity in the notions of rational acceptance and rejection.[70] Kuhn may be right that we sometimes cannot conceive of how the new paradigm could be true or see the reasons to reject an old paradigm until we have started looking at the world through the lens of the new paradigm. Nonetheless, the fundamental abilities to conceptualize phenomena in various ways and formulate reasons for them were the same both before and after the transition. Because of this fact, a successful transition can provide an "error theory," a new conception of reality that provides a rationally acceptable account of why it had seemed that the old paradigm was correct. It thus saves the appearances by showing that they were illusions, misinterpretations of the appearances by the old paradigm. Thus Newtonian physics reconceptualizes weight as a relation, and it shows why folk physics had supposed weight to be constant, namely, because it seems true from an earthbound perspective.

But it is not at all clear that such an error theory could be provided if we abandon FP.[71] Conceptions and illusions are intentional entities. So if EM is true, we can neither reconceive the phenomena nor discover that we had been harboring the illusion that there were intentional states. Nor is it clear what it can mean to have a reason to reject FP in a framework that no longer includes recognizable reasons. Are we to say that people seemed to think things although they did not? No, because there were no such things as seemings. Can we avoid this fallacy by a linguistic formulation: it is false that people acted for reasons? No, because if the new paradigm is true, then there is no such entity as the assertion that people acted for reasons that can be false. The problem is that the whole idea of an error theory is undercut by the denial of any coherent entities, such as meanings or contents, that could be false. Perhaps from the perspective of EM, FP is "not even false";[72] but this is a Pyrrhic victory if EM is necessarily "not even true."

Churchland will no doubt reply, as he frequently has, that such arguments beg the question. He may argue that from within the new paradigm, replacements will emerge for our current notions of thought, rationality, and truth. However, this position seems to me an abuse of the phrase "beg the question," since, as I argue in the next section, it is fundamental presuppositions of science itself that make EM incoherent. Since EM is promoted as a scientific program, it cannot exempt itself from criteria integral to scientific activity. Furthermore, Churchland's claim that replacements may be found remains an essentially vacuous promissory note, unless the precise relation between the replacement notions and existing notions can be specified. Not any old replacements will do; otherwise, EM is guilty of an equivocation, simply changing the subject from anything that could credibly do the same work that thought, rationality, and truth currently do. Yet, it seems antecedently likely that any replacements that we would allow would have to retain a significant amount of our current conception so that only reform, not elimination, would be in order. If Churchland insists that we must simply abandon the notions of thought, rationality, and truth, then it is enough to reply that we cannot, since we currently need them to navigate through the maze of human interaction and communication. No one will be persuaded to gouge out one's intellectual eyes without a proven replacement that will restore something resembling mental vision. Even if we allow Churchland the tactic of issuing promissory notes to ward off the threat of incoherence, his proposals remain radically premature and unmotivated.

Materialism and the Rationality of Science

These objections seem enough reason to reject EM. But it is important to avoid fixation on the specifics of that theory since it is merely a particularly pungent version of a more general program, SAR. What I wish to argue now is that the scientific materialist cannot adopt SAR in any form, because it undermines the rational practice of science. The first point I want to make is that SAR is committed to an unmotivated and uninformative approach to scientific explanation. This argument, by itself, is not conclusive, but it becomes so in conjunction with a second consideration, namely, that SAR is incompatible with the rational principles of scientific method.

The aim of SAR is a thoroughgoing evacuation of the concepts of design and intentionality from nature. Darwin had suggested a way to dispense with these concepts in the human natural environment, but he generally assumed that humans themselves designed things and had intentions. In fact, the whole force of Darwin's argument, in its historical context, presupposed this distinction between humans and their environment. Before Darwin, the British natural theologians had argued that because biological organs such as the eye exhibited an intricate structure of parts mutually adapted to perform a common function, they were more like the artifacts of an intelligent designer than the products of law or chance acting alone. To overcome this powerful line of reasoning, Darwin had to argue for a contrast between the class of actually designed objects (such as Paley's famous watch on the heath) and those that merely appeared to be designed but in fact resulted from the blind mechanisms of descent with modification and natural selection. Thus, at least at the time, the whole force of Darwin's argument rested on the presupposition that design is a legitimate concept but one whose domain of application is more restricted than had been thought.

According to SAR, however, Darwin's strategy can be applied to human beings as well, as it is by Churchland's EM. Thus, because there are no intentions and there is no designing going on, even Paley's watch on the heath is the result of entirely nonintentional, nonteleological processes, such as appropriate transformations of activation vectors. This means, however, that in explaining the existence of apparent design, there is no longer a contrast to be drawn with actual design because there is no such thing. As a result SAR undermines Darwin's original motivation for explaining the appearance of design. His motivation had been to make a distinction between the appearance of design and the reality, not to deny the existence of design *tout court*. SAR must now distance itself from Darwin's motivation and argue that design is entirely chimerical. The problem with this move, as we shall see, is that the proponent of SAR cannot dispense with design concepts in understanding the scientific method it prizes. This method involves the careful design of theories, instruments, and experiments, where the idea of these things precedes their implementation in a way that never occurs in the blind undirected processes of natural selection or in algorithmic transformations of neural-activation patterns.

Van Fraassen has convincingly argued that explanations typically get their force via a contrast between the *explanandum* and other relevant alternative possibilities.[73] Thus, when the judge asks the defendant, "Why did you rob the bank?" the judge assumes that the defendant had alternative actions, such as staying home or going to work, and wants to know why the defendant robbed the bank instead. As van Fraassen says, "In general, the contrast-class is not explicitly described because, *in context*, it is clear to all discussants what the alternatives are."[74] It may be that not all explanations have to be construed as contrastive.[75] If the question is, "Why did he fall off the cliff?" then the only "contrast" may be "not falling off the cliff," which is trivial since p necessarily excludes $\sim p$. However, the contrastive approach is best seen as a model for what it means for an explanation to be informative. According to information theory, "The amount of information associated with a state of affairs has to do . . . with the extent to which the state of affairs constitutes a reduction in the number of possibilities."[76] This elimination of alternative possibilities presupposes a contrast class, which explains why tautologous and evasive answers fail to give adequate explanations in response to "why" questions, whereas specific contingent information is generally acceptable. Thus, in response to the judge's question, if the defendant gave a reason that was similar to the famous reason given for climbing Mount Everest, "Because it was there," it would not be well taken by the judge, because that reason (or nonreason) could be equally applied for the defendant's staying at home or going to work. But the reply "Because I am homeless, unemployed, and over my head in credit-card debt" would immediately explain why the defendant did not stay at home or go to work but robbed the bank instead.

It follows from these considerations that SAR's explanations of apparent design lose the explanatory force that had been the rational motivation for Darwin's explanation. While Darwin could point to actual design as a contrast with apparent design, the proponent of SAR can draw no such distinction. As a result, when SAR offers an explanation of the appearance of design in human artifacts, the explanation, if correct, seems unmotivated and uninformative. Surely it eliminates an idea that we may have thought was possible, namely, that humans really did design the artifacts; but it does not eliminate an interesting possibility since design is not a real category, being exemplified nowhere in the universe.

The problem is that if SAR is correct, design is an illegitimate concept. In *The Critique of Pure Reason*, Kant argued that such concepts as "fate" and "fortune" lacked a "deduction," by which he meant a demonstration of the "right or the legal claim"[77] to apply the concepts literally to particulars. If SAR is correct, design is a concept without a deduction. It follows that eliminating design from scientific descriptions of human beings is on a par with eliminating unicorns from biology or Santa Claus from macroeconomics: the more we trivialize the category eliminated, the more we trivialize the elimination. Darwin's explanation was informative because it said that despite appearances, biological structures are not artifacts. SAR's account is much less informative because it says despite appearances, artifacts are not artifacts; but this statement just means that nothing is an artifact, so to learn that something is not an artifact is not to distinguish it from anything else in the universe.

At this point, a proponent of SAR might try to reply as follows.[78] Suppose that some scientist, say Boyle,[79] took the view that while planetary motion was merely mechanical, life was not. He might argue that while the planets had appeared to be guided by vital spirits, vital spirits were only a reality in living things. But then biochemistry suggests later that vital spirits do not exist at all. Boyle's project of distinguishing real from apparent vital spirits is undermined, which is an inevitable consequence, considering that there were no vital spirits in the first place. Why should not design have a similar fate? The reason that design is relevantly different from vital spirits is that it is implicated in the scientific method in a way that seems nondetachable. We can suppose that scientists are alive and rational, whether they are animated by vital spirits or not, because we have clear alternative accounts of what it means to be alive and because rationality is not tied to the vital-spirit theory.[80] However, if reason guides a scientist's investigation, theory construction, and experimentation, then it is not in the least clear what could count as an alternative to design as an account of reason's guiding role. Could we really call scientists rational if they did not design their theories so as to best fit the data and if they did not design their experiments to control for possibly interfering factors?

Again, the proponent of SAR might suggest that design is a useful fiction and that it is possible to explain why a fictitious concept seems to apply, although it never does. Certainly we can offer an explanation of why a doctored horse appears to be a unicorn or a bearded man appears to be Santa, even though there are no unicorns and there is no Santa. But we can do so only because we are contrasting the facts with a fictional account that specifies what it would be like if unicorns or Santa existed. The problem with using this strategy to defend SAR is that the ability to construct fictional accounts seems to presuppose actual design, namely, authorship. So if we say that we are contrasting the appearance of design with a fictional or conventional account of it, then we simply relocate the problem of explaining why human artifacts seem designed to the problem of explaining why human fictions seem designed. If SAR is to overcome the apparent pervasive presence of design in human activity, it must show how the concept of design arose, even though design does not exist.

But to do so is highly problematic. I pursue this issue at length in chapter 7, but for now I wish to note two major problems for the attempt to claim that design is an illegitimate concept. First, we need to ask ourselves, "What would make actual design possible?" For an artifact or action to be designed, some preexisting plan would need to exist in accordance with the artifact or action produced; furthermore, the artifact or action would need to be a causal consequence of that plan.[81] This in turn assumes that the plan is intentional, pointing in advance to a nonexistent object that completes it. Is there anything to which SAR is committed that has this property of intentionality? Yes. Since design is an empty concept, according to SAR, proponents of SAR are thus committed to the existence of concepts; however, concepts themselves are intentional. It is therefore self-defeating for SAR to claim that the concept of intentionality is empty. Indeed, it is precisely the fact that concepts are intentional that makes it possible to suggest that the concept of design might be empty: the ability to characterize a nonexistent object is one of the marks of the intentional. Concepts are also intentional in that one can conceive of *F* and not G even though all *F*'s are G's. Thus, one can conceive of an artificial heart without conceiving of a rhythmic sound maker, even though it is necessary that all artificial hearts of the sort conceived are such sound makers. But if concepts are intentional, then the prior conception of an artifact or action can constitute its plan, or "design." That notion, coupled with the reliable correspondence between such plans and the subsequent artifacts and actions produced, is strong evidence for a causal connection between the plans and the products. But if plans produce products, then those products are designed. The admission of concepts, coupled with the strong evidence that these concepts produce artifacts and actions, is a strong reason to conclude that design is a legitimate causal category and hence a strong reason to reject SAR.

Of course, a proponent of SAR might try to avoid this line of reasoning by denying the existence of concepts altogether and arguing that human cognition and action can be understood in entirely information–theoretic terms.[82] This strategy, however, does not work. As we shall see in chapter 7, practical and theoretical reasoning manifest a kind of complex specified information that is demonstrably beyond the reach of chance and necessity, whether singly or in combination. Thus, even if the very idea of concepts is rejected (a proposal whose coherence is doubtful at best), the informational characteristics of human cognition and action point to design. But if design is a legitimate category when applied to human thought and activity, it is surely possible that the category applies to nonhuman or superhuman cases as well. Once it is shown that a category is legitimate, the range of its application is clearly an empirical matter.

SAR's account of intentionality has other problems as well. According to SAR, all that nature exhibits is the appearance of intentionality. The problem is that the very idea of an appearance, even an illusory one, presupposes the reality of intentionality. To harbor the illusion that *p* is to hold the false belief that *p*, and to do so is to be in an intentional state. In addition, to insist that this only appears to be the case simply won't do, because this appearance yet again implicates intentionality and also because what we are trying to understand here is the appearance itself; thus, "*we cannot make the appearance–reality distinction because the appearance is the reality.*"[83] What needs explaining are such things as seemings, so it gets us nowhere to say that there only seem to be seemings. And even if it is logically conceivable that we have the concept of design without there being any real design, it is not conceivable that we have the conception of intentionality without conceptions, which is embarrassing for SAR because conceptions are intentional entities.

Finally, it is important to see that the concepts of design and intentionality are essential to science as a particular rational practice. Folk psychology is not just for folk; it defines (perhaps imperfectly) the practical reasoning employed by scientists every day. Scientists have projects, plans, and goals; and they learn that various ideas are either fruitful or unhelpful, true or false. Scientists think of themselves as identifying phenomena, developing hypotheses, designing experiments, and providing explanations. "But no one has shown how concepts like those of *identifying* something as a cognitive system or *hypothesizing* that cognitive states lack content have application in the absence of content."[84] How can we assert and test hypotheses if there are no such things as assertions that can be evaluated as true or false? How can we design experiments if there is no such thing as design? How can we explain anything if we do not assert anything and if others cannot understand it? The whole point of explanation is increased understanding. But to understand an explanation is to have beliefs about what the world would be like if the explanation were true. These beliefs have intentional content (they concern possible states of affairs), so understanding is intentional. And this holds even in the case of reductive explanations, which Churchland himself says are achieved by reconceiving the phenomena—that is, achieving a new understanding of it.

This is the worry expressed by C. S. Lewis in the second quote at the head of this chapter, that unlimited reductionism in science would be self-refuting since it would explain away the very notion of explanation (including reductive explanation) upon which the project of science relies. Science, as a particular exercise in practical reason, with particular goals such as explanation, cannot continue if it denies the foundations presupposed by those goals (such as understanding) or the very idea of having goals at all. What Churchland and other empiricists in the philosophy of science apparently cannot accept is that science has some nonnegotiable presuppositions (those defining the very essence of practical reasoning of which science is an application), which are not themselves vulnerable to scientific elimination.[85] It is as if a man who wants nothing more than peace and quiet starts by eliminating all the noisy distractions in his environment and ends by committing suicide because he cannot bear the sound of his own heart beat. In the quest for peace and quiet, he has eliminated the very ground for enjoying it. The general moral of this parable is that elimination itself has goals, and these set limits on how much can be coherently eliminated. If elimination were an end in itself, the best theory would assert that nothing existed, not even the scientist engaged in the elimination, which would make celebration of the achievement incompatible with what the achievement actually meant.

To remove design and intentionality from scientific practice would render it inscrutable, pointless, and practically impossible. For scientists to study and control the environment, they must think of themselves as agents in some measure detached from that environment. Scientists must make predictions and look for confirming or disconfirming evidence. They have to follow chains of argument based on this evidence and accept the most reasonable conclusions. They have to construct crucial experiments to help them decide between competing hypotheses. Every one of these activities assumes that the scientist is an agent who really has intentional states and who really implements designs. Without these concepts, there is no deduction (in Kant's sense) for the concept of scientific rationality. Scientific rationality is a highly developed application of practical rationality, and practical rationality is committed to the full reality of intentionality and design. If Churchland and proponents of other versions of SAR are to make convincing arguments, they must show how scientific rationality can survive without practical rationality or else construct an entirely new account of practical rationality itself. I see no solid reason to think that such a project is feasible, and given the current success of science with its current idea of practical reason, the onus is entirely on the proponent of SAR to prove otherwise.

Conclusion

SAR is a deeply flawed program that undermines its own mines. EM, the version of SAR proposed by Churchland, is unable to account for the ability to abstract human actions from multiple physical implementations; it cannot accommodate the fact of human subjectivity; it fails to see that intentional states would likely outlive FP, even if FP fails; and it is beset by intractable problems of coherence as soon as we seriously consider what its truth would mean—that is, if there is any such thing as truth or meaning left. More generally, the program of SAR undermines the very idea of scientific explanation. It does so by robbing explanations of the appearance of design or intentionality of a meaningful contrast with real cases and by denying the very notion of understanding that explanation presupposes. Science is a particular exercise in practical rationality, and this exercise cannot be understood, indeed cannot exist, in the absence of design and intentionality. SAR thus proposes a program of materialism that cannot claim to be scientific.

Perhaps, however, agency can be tamed by naturalizing it, by showing that though it exists, agency can be understood in purely materialist terms. That is the project of WAR, to which we turn in the next chapter.

Notes

Thanks to Del Ratzsch, for encouraging me to clarify my ideas in this chapter and respond to important objections.

1. Thomas Nagel, *The View from Nowhere* (New York: Oxford University Press, 1986), 111.
2. C. S. Lewis, *The Abolition of Man* (New York: Macmillan, 1955), 91.
3. For a good, reasonably impartial survey of the history of EM and the arguments for and against it, see the entry on "eliminativism" by Teed Rockwell in the online dictionary of philosophy of mind at www.artsci.wustl.edu/~philos/MindDict/eliminativism.html.

4. An excellent sympathetic account of FP is given by Jerry Fodor in his "The Persistence of the Attitudes," the introduction to his *Psychosemantics: The Problem of Meaning in the Philosophy of Mind* (Cambridge, Mass.: MIT Press, 1987). An excellent general collection on FP's relation to scientific psychology is John D. Greenwood, ed., *The Future of Folk Psychology* (Cambridge: Cambridge University Press, 1991).

5. What counts as "rational" in folk psychology has nothing to do with whether or not the action would be pursued by an ideal agent or is the best game–theoretic choice. It simply means that the action is intelligible in light of the agent's desires and beliefs about how to satisfy those desires, and so the notion of rationality is in that sense relative, not absolute.

6. See the BBC science site www.bbc.co.uk/dna/h2g2/A597152.

7. See, for example, "Activation Vectors vs. Propositional Attitudes: How the *Brain* Represents Reality," chapter 4 in the Churchlands's *On the Contrary*.

8. Churchland favors the idea of a prototype, where objects are judged to be more or less similar to certain exemplars of a class (e.g., a robin might be an exemplar for the class of birds), and thus there need be no sharp divide between members and nonmembers of a class.

9. Paul Pietroski defends the Fregean thesis that to believe that *P* is to believe the thought expressed by "*P*" in its appropriate context. See his *Causing Actions* (Oxford: Oxford University Press, 2000), 57. It is worth noting that this account does not imply that the believer is related to a sentence or other languagelike entities, as Churchland often suggests.

10. Wilfrid Sellars, "Empiricism and the Philosophy of Mind," in *The Foundations of Science and the Concepts of Psychology and Psychoanalysis*, ed. H. Feigl and M. Scriven, Minnesota Studies in the Philosophy of Science, vol. 1 (Minneapolis: University of Minnesota Press, 1956).

11. Paul Churchland, "Folk Psychology," in the Churchlands's *On the Contrary*, 8.

12. Paul Churchland, "Eliminative Materialism and the Propositional Attitudes," *Journal of Philosophy* 78, no. 2 (February 1981): 67–90, 73–74.

13. It seems plausible that babies desire to be fed without their being aware that that is what they desire and certainly without their being able to express the desire linguistically.

14. For example, see Paul Churchland's *The Engine of Reason, the Seat of the Soul: A Philosophical Journey into the Brain* (Cambridge, Mass.: MIT Press, 1995).

15. Paul Churchland, "Eliminative Materialism and the Propositional Attitudes," 75.

16. Paul Churchland, "Folk Psychology," 8.

17. Paul Churchland, "Folk Psychology," 8.

18. Churchland uses considerable rhetoric to suggest that FP has failed to expand its domain of application. But it would be a clear fallacy to argue that, say, deductive and inductive logics are stagnant research programs because so few important sciences talk about them. Obviously, deductive logic and inductive logic have been of huge importance to science by helping scientists develop more accurate and testable

theories about other things. Why should not FP's contribution to science, as the basis for the practical rationality of the scientist, be defended in just the same way? For more on this, see the section entitled Materialism and the Rationality of Science.

19. Churchland, however, is wrong to assume that commitment to propositional attitudes implies that our minds or brains manipulate sentencelike entities. All that is required, as Pietroski points out, is that our minds are related to thoughts expressible by sentencelike entities. See endnote 10 in this chapter.

20. Paul Churchland, "Folk Psychology," 14.

21. Paul Churchland, "Folk Psychology," 15. It is unfortunate that in describing this "alternative" model, Churchland himself continues to make use of intentional states such as recognition and anticipation, most naturally expressed by propositional attitudes.

22. Although, to be fair, Churchland sometimes waxes more modestly and suggests only that it is unlikely that FP will survive; however, because it is an empirical matter and all the relevant data are not yet in, this cannot be known with certainty.

23. See Hilary Putnam, *Representation and Reality* (Cambridge, Mass.: MIT Press, 1988); and Daniel Dennett, "Two Contrasts: Folk Craft versus Folk Science, and Belief versus Opinion," in *The Future of Folk Psychology: Intentionality and Cognitive Science*, ed. John D. Greenwood (Cambridge: Cambridge University Press, 1991), 135–48.

24. John D. Greenwood, "Reasons to Believe," in *The Future of Folk Psychology*, 88.

25. John Heil, "Being Indiscrete," in *The Future of Folk Psychology*, 130.

26. In the face of this challenge it can be urged that the retention of agency as a significant scientific category means defending a program like ID, which rehabilitates agency in natural science and demonstrates its irreducibility to physicalist categories.

27. Of course, it is highly debatable whether FP needs confirmation from neuroscience. If FP has no real implications for neuroscience, then of course it cannot be confirmed by it; but that does not make FP false. Presumably, the principles of carpentry have no implications for neuroscience either, but that does not show that those principles are unsound.

28. It is, of course, always possible to look at neurophysiology from an intentional stance. However, if that means taking Dennett's real patterns seriously, it is, to that extent, valid reason to take the intentional states of folk psychology seriously.

29. Laplace had famously claimed that given complete physical knowledge of the universe at any given time, one could predict all of its subsequent states. Quantum physics apparently shows this is impossible, but we will suppose that the Martians have the closest approximation possible to Laplacean knowledge.

30. Dennett, "True Believers," in his *The Intentional Stance*, 13–35, 26.

31. Dennett, "True Believers," in his *The Intentional Stance*, 26.

32. This is why I use the example of economics and not Greenwood's example of social psychology. While I am very sympathetic to Greenwood's argument, in the case of social psychology, Churchland can counter that it grew out of FP; so perhaps it needs to be replaced along with FP.

33. Likewise, in our earlier example of FP, Thatchwood has good reason to think that Hagarth-Smythe will murder again, even though it is unknown which of an indefinite number of techniques Hagarth-Smythe will employ.

34. Intervening layers being equal, similar input-activation patterns lead to similar output-activation patterns, but this means similar movements. The problem is that the same action can be performed by vastly different sequences of movement.

35. It is interesting that while many in the philosophy of mind are convinced by such arguments that mental states are indispensable, few philosophers of biology will accept an equally strong parallel argument for design, which calls attention to the fact that many species have independently converged on the same type of solution to an environmental problem, even though the physical causes must be different. Since the same kind of adaptation can be built out of quite different biochemical materials, the only common feature of the convergent adaptations is that they solve the same problem, which points to design just as strongly as the wide variety of movements implementing a common action points to a common intention.

36. Some maintain that only a subvenient physical base has causal efficacy so that the content of the intentional states is epiphenomenal. This view is consistent with either an instrumentalist account of intentional states as a useful fiction (the early Dennett) or the claim that FP provides a noncausal rational explanation (see, for example, Grant Gillett's "Actions, Causes and Mental Ascriptions," n. 55, ch. 1). Others, such as Fred Dretske, argue that content has a higher-level role in structuring the relation between brain states and behavior (see Dretske's *Explaining Behavior: Reasons in a World of Causes* [Cambridge, Mass.: MIT Press, 1988]). Yet others argue that mental causation is simply a fact because they are unabashed substance or property dualists. For a recent defense of the former, see Charles Taliafferro, "Naturalism and the Mind" in *Naturalism: A Critical Analysis*, eds. William Lane Craig and J. P. Moreland (London: Routledge, 2000), 133–55; for a recent defense of the latter, see Paul M. Pietroski, *Causing Actions* (Oxford: Oxford University Press, 2000).

37. Searle, *The Rediscovery of the Mind*, 116–24.

38. Searle, *The Rediscovery of the Mind*, 124.

39. Searle, *The Rediscovery of the Mind*, 122.

40. Churchland, "Betty Crocker's Theory of Consciousness," in the Churchlands's *On the Contrary*, 113–22.

41. Churchland, "Betty Crocker's Theory of Consciousness," 117.

42. Searle, *The Rediscovery of the Mind*, 117.

43. Churchland, "Betty Crocker's Theory of Consciousness," 118.

44. Churchland, "Betty Crocker's Theory of Consciousness," 118.

45. See especially Nagel's *The View from Nowhere*, 25–27.

46. If one wants to make it a definitional truth that objectivity encompasses everything, one will have to say that objectivity includes subjectivity, trivializing the distinction.

47. Indeed Churchland himself relies on this distinction when he talks of FP as an internalized theory providing a certain conception of reality, which makes it appear

that there are intentional states. The fact is that reductionist theses cannot even be coherently stated without supposing that there is a coherent appearance–reality distinction both before and after the reduction is effected. Thus, reductionists typically show that what appears to be an independent reality is not.

48. Paul Churchland, *Matter and Consciousness* (Cambridge, Mass.: MIT Press, 1984), 56.

49. Churchland, "Folk Psychology," 8.

50. Greenwood gives another good example. At one time it was widely believed that marsh vapor caused malaria. The discovery that marsh vapor plays no such causal role did not convince anyone that marsh vapor does not exist. See Greenwood, "Reasons to Believe," in his collection *The Future of Folk Psychology*, 77.

51. Greenwood cites the work of A. M. Leslie, "Some Implications of Pretense for Mechanisms Underlying the Child's Theory of Mind," in *Developing Theories of Mind*, ed. J. W. Astington, P. L. Harris, and D. R. Olson (Cambridge: Cambridge University Press, 1988).

52. Greenwood, "Reasons to Believe," in his collection *The Future of Folk Psychology*, 81.

53. Searle, *The Rediscovery of the Mind*, 61.

54. See H. P. Grice, "Meaning," *Philosophical Review* 66: 377–88; and J. R. Searle, *Intentionality: An Essay in the Philosophy of Mind* (Cambridge: Cambridge University Press, 1983).

55. As we have just seen, the alternative view, that linguistic behavior preceded commitment to intentional states, is incoherent.

56. Lynne Rudder Baker, *Saving Belief: A Critique of Physicalism* (Princeton, N.J.: Princeton University Press, 1987), 134.

57. See Baker, *Saving Belief*, 135–38.

58. See Baker, *Saving Belief*, 138–42.

59. See Baker, *Saving Belief*, 143–47.

60. It is tempting to suggest that Churchland is advancing a kind of mysticism. Believers in FP are in the vale of maya, or illusion, which makes factitious distinctions between true and false opinions; but the enlightened ones, those who have eliminated all illusion, see that all is one. The obvious objection is that the concepts of illusion and "seeing that *p*" are intentional. No doubt, however, these are merely the steps of the stepladder the mystic uses in ascent but no longer requires. For more musings on these lines, see John Heil's "Being Indiscrete," in *The Future of Folk Psychology*, 121–22.

61. Churchland, "Evaluating Our Self-Conception," in the Churchlands's *On the Contrary*, 28–29.

62. Churchland, "Evaluating Our Self-Conception," 29–30.

63. Churchland, "Evaluating Our Self-Conception," 29.

64. The leftmost vertical bar indicates the main derivation, under the main premises of the argument from line *1* to line *k*. To the right is another vertical bar indicating the subderivation, under the working assumption at line *n* is to be refuted by reductio.

65. It may be objected that this extreme example is especially paradoxical because we would all agree that reductio has the status of logical necessity, whereas we might disagree about the status of the practical rationality prescribed by FP. But it is easy to construct other parallel cases that make essentially the same point. For example, suppose that I am a postmodern extremist, and I write a lengthy book in French claiming to show by reductio that language is unable to communicate information. Readers will be forgiven for not accepting my conclusion because they realize that if they do accept it, I must have communicated information to them and so must be mistaken in my thesis. Thanks to Del Ratzsch for this example.

66. Baker, *Saving Belief*, fn. 3, 136.

67. Translation: Churchland would need to appropriately activate our neurons in such a way that we do the counterpart of what we now call "accepting" and that the replacement notion does the same kind of job (only better) that acceptance currently does in logical reasoning.

68. Churchland, "Evaluating Our Self-Conception," 29.

69. If Churchland's point is merely that gruntal attitudes might not properly be propositional attitudes, then he needs to give an account of contemplation that is not equivalent to attributing such an attitude. But I think this would be hairsplitting, since, to the extent one can eliminate propositions, one can also give an account of FP without propositional attitudes. It is possible to hold that while statements involving that clauses are used to attribute intentional states, these states are not literally to be understood as relations to propositions. The crucial point is that these states have content, not whether propositions are exactly the right way to express that content.

70. In fact, Churchland himself tacitly assumes such continuity by describing reduction as a process of reconceptualization, which assumes that there are such things as concepts before and after the reduction (see Churchland, "Intertheoretic Reduction," 69–70). He also assumes such continuity by supposing that available to him after the transition will be a concept of coherence, which will allow us to see that the incoherence of abandoning FP is only apparent (see Churchland, "Evaluating Our Self-Conception," 29). Yet, what other than something very like propositions and our attitudes to them, the very things denied by EM, can be either coherent or incoherent?

71. See Baker, *Saving Belief*, 140.

72. Dembski recounts the case of physicist Wolfgang Pauli, who disparaged another scientist's work as "not even false," implying that it was so confused that it had not achieved the dignity of error. See Dembski, *Intelligent Design*, 70.

73. See Bas C. van Fraassen, "The Pragmatics of Explanation," chapter 5 in his *The Scientific Image* (Oxford: Clarendon Press, 1980), especially section 2.8 (Why Questions), 126–30.

74. Van Fraassen, "The Pragmatics of Explanation," 128.

75. So David Hillel-Ruben argues in his *Explaining Explanation* (London: Routledge, 1990), 39–44. While Ruben exposes important technical flaws (such as intentional fal-

lacies) in some formulations of the contrastive view, they do not repudiate the basic insight of information theory that maximal information is maximal elimination of alternatives.

76. Fred Dretske, *Knowledge and the Flow of Information* (Cambridge, Mass.: MIT Press, 1981), 8.

77. Immanuel Kant, *The Critique of Pure Reason*, trans. Norman Kemp Smith (London: Macmillan, 1982), 120.

78. Thanks to Del Ratzsch for this example.

79. This example is not intended as Boyle scholarship and need not reflect his considered views. While generally hostile to "active principles," Boyle did hold that such entities were needed to explain the development of living things. See Nancy Pearcey and Charles Thaxton, *The Soul of Science* (Wheaton, Ill.: Crossway Books, 1994), 88.

80. True, as Churchland would stress, in Boyle's time, there may not have seemed to be a serious alternative to vital spirits. But the most that the proponent of SAR can argue on this basis is that were we to have a serious alternative to design as a way of understanding human rationality, design would come under serious pressure. But this is no more than a promissory note, given the entrenchment of design in our understanding of scientific method.

81. Strictly speaking, "designs" in some nominal sense could really exist and yet be epiphenomenal. But our interest is in design as a legitimate scientific category, and this requires that it is also a causal notion.

82. This means that the challenge to design can no longer be posed in terms of the suggestion that design is an empty concept, since there are no concepts. Perhaps, however, it can be claimed that design constructs of some other kind do not attribute a real property to their objects.

83. Searle, *The Rediscovery of the Mind*, 122.

84. Baker, *Saving Belief*, 136.

85. This is not to be dogmatic and deny the possibility of reform. Indeed, symbolic logic has surpassed (without simply replacing) the notions of reason employed in ordinary life.

CHAPTER THREE

~

Weak Agent Reductionism: Science and the Rationality of Materialism

> Our intentionality is derived from the intentionality of our "selfish" genes! . . . But then who or what does the designing? Mother Nature, of course, or more literally, the long, slow process of evolution by natural selection.[1]

> There isn't any Mother Nature, so it can't be that we are her children or her artifacts, or that our intentionality derives from hers.[2]

Weak Agent Reductionism

Even among philosophical naturalists, strong agent reductionism (SAR) is widely regarded as an implausible view. More popular are versions of what I am calling weak agent reductionism (WAR). WAR agrees with SAR that the appearance of agency in our natural environment can be explained away. But WAR insists that the agency of humans (and perhaps some other creatures) is real because, as we have seen, agency is indispensable to human practical rationality. Yet if human agency cannot be eliminated, WAR must show how it can be naturalized. According to WAR, human agency is either simply an odd, irreducible aspect of nature, or else it can be given a conservative or reforming reduction to physical states or processes, either of the present state of the agent (a synchronic reduction) or of its causal ancestors (a diachronic reduction).

Many proposals, sometimes overlapping, may implement WAR's program. Of the physicalist programs, the most popular has been functionalism. I begin by noting some standard objections to functionalism that also cast doubt on physicalism in general. This will motivate Dennett's alternative approach as a more promising proposal for naturalizing agency. Next, I clarify Dennett's notion of the intentional stance and his claim that our intentionality derives from Mother Nature. I argue that Dennett fails to explain the origin of intentionality. Either Mother Nature is natural selection, in which case she is demonstrably insufficient to generate intentionality; or Mother Nature is attributed the same kind of intentionality she is invoked to explain, precluding a materialist explanation of intentionality. Finally, we step back from the specifics of Dennett's account and argue that WAR in general is unable to close the door to nonmaterial agency.

The Case for Naturalized Agency

Until fairly recently, philosophy of mind has been dominated by physicalism. Physicalism has been formulated in a variety of ways, but the most common versions assert that our ontology includes only those entities denoted by terms in our best physical theories.[3] It is typically assumed that such sciences as physics and chemistry are "physical" while psychology, sociology, and economics are problematic. Biology is in an uneasy middle position since the composition of biological structures is reducible to chemistry, but their functional organization seems to resist such reduction. Nonetheless, it is frequently assumed that biology is (or is reducible to) physical science, and thus a natural approach for physicalists is to argue that psychology reduces to biology. Reduction, in the sense of naturalization, implies either a conservative reduction, that is, mental states are identified with physical states, or a reforming reduction, that a genuinely scientific theory will correct the folk-psychological account of the mental while retaining its legitimate insights.

According to some, this reduction is a synchronic reduction of mental states to neurophysiological states. For example, Jaegwon Kim has persuasively argued that functionalism is such a program, even though the received view claims that functionalism is a nonreductive approach.[4] The received view is defended by the multiple realization argument (MRA). According to MRA, the same mental state might be realized in different ways in different species; in nonliving automata; and perhaps even in the same individual, depending on maturation, degeneration, or brain damage. Consequently, the attempt by the identity theory to type-identify mental and physical states seems unlikely to succeed. Even if this is true, it is not a valid reason to claim that functionalism is nonreductive. Multiple realization occurs in physics—for example, temperature is realized in different ways in gases, solids, and vacuums—but this does not threaten the reduction of temperature. It simply shows that the reduction must be relativized to a certain domain: temperature of a gas, temperature of a solid, and so on.

Kim defends a similar strategy for mental states.[5] According to Kim, we can understand a putative mental state as a second-order property M that specifies a causal role. This role can be played by a number of different first-order physical realizing properties: $P_1, P_2, \ldots, P_k$. Physicalists only recognize the existence of properties if they play a distinct causal role. It is clear, however, that the causal role of M is exhausted by the causal roles of $P_1, P_2, \ldots, P_k$. It is also implausible to claim that in addition to such properties is the disjunctive property, $P_1 \vee P_2 \vee \ldots \vee P_k$, since in any particular case it is simply P_1 or P_2 or . . . P_k that does the work. Thus, it is best to view M not as a property in its own right but merely as an abstract concept or description that picks out P_1, P_2, . . . P_k. To put it another way, M is not another property in addition to P_1, $P_2, \ldots, P_k$. That would be a Rylean "category mistake," like supposing that Oxford University was something else in addition to all of its colleges.[6]

However, as we will soon see, the trouble with functionalism is that either it does not work or it trivializes physicalism. By noting the inadequacies of functionalism, I also hope to bring out some general inadequacies of orthodox physicalism and motivate Dennett's alternative as a more promising approach and therefore a more worthy opponent for the rest of the chapter.

One objection to functionalism is straightforwardly metaphysical. If M is a mental description multiply realized by physical properties $P_1, P_2, \ldots, P_k$, then in virtue of what feature, common to all the P_i, do they satisfy the description M?[7] It obviously won't do to say that all of the P_i share the property of being describable by M since this is not a real property but only a Cambridge property and since we want to know in virtue of what real property this Cambridge property is instantiated. The physicalists cannot answer with a common mental property since that reintroduces what they aimed to reduce; therefore, they must answer with a common physical property. Given the extent of possible multiple realization, the only plausible candidate would be a functional physical property, call it P_F. If this is right, then Kim is mistaken in claiming that we have really explained anything by reducing M to $P_1, P_2, \ldots, P_k$. The only reason that this reduction seems to work is because anything that instantiates one of the P_i thereby instantiates P_F, and it would be simpler and more illuminating to say that M is reducible to P_F.

For this to work, however, we must assume that functions can be understood in purely physical terms. The problem is that if we do this, important mental qualities are left out, while if we keep those qualities in, the functions are not physical in any sense that makes physicalism an interesting thesis.

One of the best-known attempts to specify functions in purely physical terms is the program of *strong artificial intelligence*.[8] On this view, "different material structures can be mentally equivalent if they are different hardware implementations of the same computer program." Thus "the mind just is a computer program and the brain is just one of the indefinite range of different computer hardwares (or 'wetwares') that can have a mind."[9] Although it can be debated,[10] let us allow that the concept of a program provides a legitimate physical notion of a function. The problem is that physical systems can implement such programs without essential features of a mind.

Many of the problems center on subjectivity and seem to apply to all versions of functionalism and, in fact, all current versions of physicalism. From a genuinely physicalist point of view, the "function" of a computer program is a set of impersonal transitions between physically specified elements, mapping inputs to internal states of the computer and mapping those states to output. Since these transitions can be fully specified in impersonal terms and do not depend in any way on a point of view, there is no reason to think that the computer has any sensations or consciousness. The "raw feels," or qualia of subjective experience, may be absent[11] or inverted;[12] and the mappings will be unchanged. So the execution of the function has nothing to do with the system's having such experience.[13] Not only that, even supposing that the computer program supplies reasons for action, there is no evidence that they are the personal reasons of an agent. A computer can implement all sorts of programs to produce apparently intelligent behavior without the kind of self-representation that is needed to convert a reason to do action A to *my* reason to do A. For that matter, there is no reason to think that there is a self so that all and only the representations internal to the computer belong to it as a coherent unity, rather than merely occurring in the computer as independent atoms. All of these problems reflect the fact that the functions of a computer program can be fully specified in impersonal terms, whereas agents have particular points of view, personal experiences, and personal reasons for action.

More fundamentally, there is no convincing evidence that the computer has even impersonal representations or reasons, if indeed such things exist.[14] While some philosophers seem to miss the point of his argument, Searle showed this point definitively a long time ago.[15] Even if impersonal representation or intentionality does exist, it cannot exist without content. Representations are about the world and can be true or false of it. The problem is that no matter how intelligent a computer's behavior may seem, there is no reason to think representational content plays any role in the production of that behavior. In Searle's famous "Chinese Room" example, a person is locked in a room and receives Chinese symbols that are questions in Chinese. He also has a mapping program that could be implemented rather tediously as a lookup book. Suppose an input symbol (a question) is always on a left-hand page, and the corresponding output symbol (the answer) is on the facing right-hand page. The person in the room is then able to give the right Chinese answer to the question. But the person understands no Chinese, and the mapping between input and output is accomplished entirely by matching the patterns, or "syntax," of the symbols so that their meaning or content plays no role.[16]

This, in fact, is exactly how computer programs, even highly sophisticated ones, generate output. A certain input pattern is manipulated in virtue of its form to produce a certain output pattern.[17] Indeed, that this is what happens in connectionist models of the brain is precisely the reason an eliminativist may deny the existence of mental contents: vector-to-vector transformations do not at all depend on the content of activation vectors. With a computer, the illusion of understanding is generated because the program gets what we understand as the right answer for what we understand as the question.[18] But there is no reason to think that the computer requires any understanding or any form of representation to implement the program. Nor is there any reason to say that the program represents anything, except in the derived sense that the programmer can treat it as a set of linguistic instructions that convey his intentions.[19]

These failures of functionalism all seem to stem from the attempt to understand personal agency from the impersonal view of physics. For a function to be purely physical, it must be reducible to blind mappings between inputs, internal states, and outputs; and neither the mappings nor the items mapped have any intrinsic meaning. Consequently, there is no room for intentionality. As Searle notes, this is not a peculiarity of functionalism but traces to a fundamental weakness of physicalism, its insensitivity to the normative:

> A symptom that something is radically wrong with the project is that intentional notions are inherently normative. They set standards of truth, rationality, consistency, etc., and there is no way that these standards can be intrinsic to a system consisting entirely of brute, blind, nonintentional causal relations.[20]

To capture normative notions, an enriched notion of function is required that goes beyond a mere mapping and captures the ideas of goals and purposes. Some have claimed that such a notion derives from the natural teleology of evolution.[21] We will examine this idea in later sections, but for now simply note that it is unpromising, as the whole point of Darwinism is that random mutations and natural selection have no purpose. At best, Darwinism could explain why there appear to be functions in the rich sense, although there are not. And this would not even explain the appearance of intentionality since, as we have already seen, the existence of such appearances presupposes that intentionality is real and not merely an appearance.[22]

Others have opted for a causal theory of representation that aims to show how an internal state can represent a certain property in virtue of its causal history.[23] This is a worthwhile project, but it is beset by serious problems. One is Fodor's disjunction problem, which points out that if representations mean whatever causes them, then there are no false representations.[24] If someone's "horse" representations are caused by horses and some cows, how do we argue in the latter case that these are misrepresentations of cows rather than accurate representations of cows or horses? The most plausible solution invokes the idea that horses are what cause the representation when the cognitive system is properly functioning. However, the idea of proper functioning invokes normative notions that govern what the system is supposed to do and are every bit as problematic for physicalism as intentionality.[25] Indeed, in sentences of the form "X is supposed to do Y," Y is itself an intentional context,[26] and it arguably implies an agent since it makes sense to ask "Supposed by whom?" If so, such norms presuppose intentionality and agency and thus cannot appear in a reductive explanation of them. Furthermore, even if a physicalist account of representation works, the problem is that the representations are still impersonal and atomistic.[27] There is no reason to think that the representations belong to an agent or give the agent reasons for action. In addition, the fact that a given property is represented does not explain why it is a goal that someone wants to achieve (a desire) or something held to be true (a belief). Causal theories remain wedded to an impersonal physical view, and they seem incapable of accounting for the personal perspectives and reasons of an agent.

The final option for the physicalist is to accept the reality of full-blooded functional notions for what it is. The problem is that by invoking goals and purposes not present in blind mechanistic processes, such functions are just as problematic for physicalism as is intentionality. If the physicalists simply declare full-blooded functions "physical," they may as well do the same for intentionality. But if they do so, physicalism is a trivial thesis.[28] If the physical just means whatever we have good reason to think exists, then there are strong arguments for saying that God is physical, namely, the best arguments for saying that He exists. Physicalists who help themselves to a full-bodied notion of function or intentionality without supplying a Kantian deduction of these categories are enjoying all the advantages of theft over honest toil.

The Intentional Stance

A promising way out of these problems is to drop the idea that the only model for scientific explanation is the one supplied by the physical sciences. Might there not be distinct but equally legitimate explanatory stances that uncover different dimensions of reality? Ironically,[29] it is the devout materialist Daniel Dennett who has most consistently and persuasively argued for the affirmative.

According to Dennett, the explanatory approach of physicalism is only one of three possible stances. Physicalists recognize only the *physical stance*, which bases its account of a system's behavior on "its physical constitution . . . and the physical nature of the impingements on it."[30] However, these gory physical details are often of no interest; perhaps we want to know what a computer program does (e.g., graphing a spreadsheet) but not how it does it. In this case, it is better to adopt the *design stance*: "one ignores the actual (possibly messy) details of the physical constitution of an object, and, on the assumption that it has a certain design, predicts that it will behave *as it is designed to behave*."[31] The design stance is surprisingly efficient and successful, so long as there is no design flaw or hardware malfunction, which can only be explained by dropping back down to the physical stance.[32] Simple, designed objects, however, typically implement the goals of their designers and users, not their own. When we turn to more complex entities that appear to act for their own reasons, it is helpful to adopt the *intentional stance*: "first you decide to treat the object whose behavior is to be predicted as a rational agent; then you figure out what beliefs the agent ought to have, given its place in the world and its purpose."[33] Attributions of intentional states to an entity thus depend on norms of rationality that are invisible from the physical stance.

At first sight, this attribution of intentional states is problematic. If intentional states really exist without reservation, it seems Dennett cannot claim to be a materialist without trivializing materialism. If they do not exist and our intentional talk is a useful fiction, then Dennett is merely wrapping eliminative materialism in the mantle of instrumentalism. Neither of these interpretations of Dennett is charitable, and he has made it clear over the years that he has something else in mind.

As we saw in the last chapter, Dennett thinks that there are real patterns in human behavior that are visible from, and only from, the intentional stance. Like Churchland, Dennett is an empiricist who believes that we should accept in our ontology what our successful scientific theories say exist—at least until those theories are superseded. Since the intentional stance is successful and lacks a credible alternative, Dennett is willing to conclude that intentional states exist. However, Dennett does not think beliefs and desires are of the same unproblematic category as *concreta*, like tables and chairs. Following Reichenbach, Dennett distinguishes "*illata*—posited theoretical entities—and *abstracta*—calculation-bound entities or logical constructs."[34] While Churchland thinks of intentional states as the illata of folk psychology, Dennett claims that beliefs and desires are abstracta, on a par with centers of gravity, the equator, or ideal gases. The abstracta of the intentional stance are not instrumentalist fictions, but unlike concreta, "they figure in explanans incorporating certain idealizations made in the context of explanatory practice."[35] The idealizations capture norms of rationality that may not obtain due, for example, to an individual's fatigue, self-deception, or brain malfunction. Even when they do obtain, these norms may not succeed in uniquely identifying an agent's intentional states. Beliefs and desires are attributed as part of a pattern of mental states that provides an overall interpretation of an agent's behavior. Given the aim of optimizing rationality, there may be times when a number of different interpretations fit the facts so that it is indeterminate whether the agent really has a particular belief or desire.

> We can be sure in advance that no intentional interpretation will work to perfection, and it may be that two rival schemes are about equally good, and better than any others that we can devise.[36]

This position does not, however, threaten the reality of intentional states. Rather it shows that they cannot simply be identified with concrete material items in the way physicalists hope. As Viger argues, Dennett can be read as a "small 'r' realist" who "questions the intuition that intentional states and properties are real only if they can be identified with something physical by providing examples of abstracta for which we have no comparable intuition."[37] However, Dennett maintains that the reason for thinking intentional states exist is not that they are abstracta (some abstracta may be fictions) but that the intentional stance is predictively successful.

As a result, Dennett thinks that intentional states are no more problematic than other theoretical entities, and he feels no pressure to provide a synchronic reduction of intentional states to something else. However, Dennett realizes that this provides no explanation of the origin of intentionality. For that, some kind of diachronic account is required, and Dennett proposes that human intentionality traces to natural selection.[38]

Dennett works up to his evolutionary account by way of an analogy with an artifact, a "2-bitser" vending machine designed to accept only U.S. quarters. According to Dennett, this is a classic case of derived intentionality.

When we ascribe the intentional state of accepting a quarter to the 2-bitser, we do not mean that it originated such a state; all we mean is that the machine fulfills the human intention of accepting quarters. Just as speakers of French can use "J'ai grand soif" to mean "I am very thirsty,"[39] so human designers can use some physical state of the machine to mean "accepting a quarter." Dennett then points out that the 2-bitser could be appropriated by Panamanians whose intention is to use the machine to accept the physically similar balboas. This analogy illustrates the obvious fact that derived intentionality is not a constant, since the same physical structure can be used to fulfill different intentions. Consider the difficult cases beyond this example: First, two or more parties may simultaneously have conflicting intentions for an artifact; for instance, a letter opener is appropriated as a murder weapon. Second, no one may have any continuing intention for an artifact; for example, the forlorn 2-bitser is consigned to a landfill even though its coin-recognition unit still works. These are difficult cases for what Dennett calls "artifact hermeneutics," our reading of an artifact's function, because either the meaning has to be relativized to avoid inconsistency or it is no longer clearly defined.

Dennett realizes that the 2-bitser is too simple to be comparable to a human being, so he moves a little closer to home by considering a machine designed to keep a hibernating human occupant alive, a sort of survival robot. He imagines that the robot has sensory systems, the ability to "anticipate" danger, "find" energy, and "cooperate" or "compete" with other similar robots. Since the robot may encounter situations unforeseen by its designer, it must "be capable of deriving its own subsidiary goals from its assessment of its current state and the import of that state for its ultimate goal." As a result, the "robot may embark on actions antithetical to [the designer's] purposes."[40] The example is intended to force a choice. If we say that derived intentionality must conform to the intentions of an artifact's designer, then its novel behaviors have no intentionality at all; there is only the "as if" intentionality of simulation. However, if one, like Dennett, thinks it is arbitrary to deny intentionality to these novel behaviors, the example shows that derived intentionality can diverge from the designer's intentions.

Having thus prepared us, Dennett drops his bombshell. Following Richard Dawkins,[41] Dennett suggests that we are simply survival machines, designed to preserve our genes. Suppose this is correct: if we say that the robotic survival machine has mere "as if" intentionality, we will have to say the same thing about ourselves. This is intolerable (and Dennett has agreed that we really do have intentionality), so apparently we must conclude that real intentionality can be derived and that derived intentionality can transcend the intentions of the designer.[42]

But who is the designer in the case of human beings? Dennett admits that genes are too stupid to design anything. In fact,

> they do not do designing themselves; they are merely the beneficiaries of the design process. But then who or what does the designing? Mother Nature, of course, or more literally, the long, slow process of evolution by natural selection.[43]

Dennett's idea is that natural selection mirrors the mind in that it seemingly makes choices and improves its products by weeding out failures, yet it does so without any representations or foresight. Further, since natural selection may reuse one structure for a different function, there will sometimes be indeterminacy about what a structure is for, just as in the case of the 2-bitser. This potential for indeterminacy is inherited by our intentional states.[44] Despite these limitations, Mother Nature is mindlike enough that it is profitable to apply the intentional stance. To some degree, this is a practical necessity resulting from the limitations of physical stance accounts of biological function. "Pending completion of our mechanical knowledge, we need the intentional characterization of biology to keep track of what we are trying to explain."[45] More than that, without the intentional stance, "We would miss the pattern that was there, the pattern that permits prediction and counterfactuals."[46] Dennett agrees with Ruth Millikan[47] that the intentional stance not only helps us to identify biological function, it also determines the content of folk-psychological attributions, for "it is only relative to . . . design 'choices' or evolution-'endorsed' purposes . . . that we can identify behaviors, actions, perceptions, beliefs, or any other categories of folk psychology."[48]

In summary, according to Dennett, human intentionality is real: the intentional stance enables us to detect "real patterns." But it is not original, for "we must recognize that it is derived from the intentionality of natural selection, which is just as real."[49] Since our intentionality is derived, we do not have some privileged authorial knowledge of what we really mean, and sometimes, as in the cases of the 2-bitser and robotic survival machine, there may be no clear fact of the matter.

Dennett Denied

There is much to admire in Dennett's account and with which I have no quarrel.

I agree with his distinct explanatory stances. I also agree with the further consequence that much of physicalist philosophy of mind has been misdirected because it has either denied the facts about intentionality or distorted them by forced assimilation to an alien physical-stance perspective. What is more, I even agree that human intentionality is derived, and that it is sometimes afflicted by indeterminacy, although for quite different reasons than Dennett's.[50] Nonetheless, one has several powerful reasons for rejecting Dennett's overall account, especially his claim that human intentionality derives from natural selection.

Objection 1: The Abstraction Problem Revisited

Dennett claims that the real patterns in human behavior visible from the intentional stance show that humans have intentional states. He also presupposes real patterns in the function of biological structures that are visible from the intentional stance. So why does he not conclude that these structures are the product of an intelligent designer with intentional states?

While Dennett thinks of intentional states as abstracta, he thinks they are real because they really explain and predict the patterns of human action. He also thinks that the intentional stance is needed to see biological structures as artifacts, which supports the adaptionist program of reverse engineering. Yet he does not conclude, by parity of reasoning, that these biological structures are best explained by an agent with intentional states. Why not?

Dennett gives two reasons for rejecting the parallel argument. First, he claims that in the case of natural selection, there is only an "illusion of intelligence" because "evolution may well have tried all the 'stupid moves' in addition to the 'smart moves,' but the stupid moves, being failures, disappeared from view. All we see is the unbroken string of triumphs."[51] The second reason is a prelude to Dennett's later distinction between skyhooks and cranes.[52] To have intentionality simply appear from nowhere would be a nonexplanatory skyhook. But to expect it to appear in full form as the immediate product of some natural process would be "greedy" or "precipice" reductionism.[53] Instead, Dennett's idea is that natural selection has a sort of proto-intentionality without representations—let us call it, in honor of its inventor, "Dennettionality." When natural selection finally produces systems that are able to represent states of affairs, this crane moves Dennettionality closer to the full-bloodied intentionality of folk psychology. Although Dennett says that the intentionality of natural selection is just as "real" as human intentionality, he really seems to recognize several levels of intentionality where each higher level is supported by the chance emergence of a crane:

Darwinian creatures, whose behavioral repertoire is fixed by natural selection;

Skinnerian creatures, who can learn new behaviors in their own lifetime by operant conditioning;

Popperian creatures, who have an internal model of their outer environment and are thus able to predict the consequences of a behavior without running the risk of actually trying it; and

Gregorian creatures, whose thought and behavior can be modified and extended by the use of artifacts, such as scissors or the ultimate "mindtools," words.[54]

This account is as ingenious as it is implausible. From the fact that natural selection has no representations, what Dennett should conclude is that it does not exhibit intentionality at all. By definition, intentionality may be directed to nonexistent states of affairs, as when a child longs for the arrival of Santa Claus.[55] But nothing can be literally related to a nonexistent state of affairs, and so any plausible account of intentionality holds that a child has a representation whose content is that Santa will arrive. This representation helps because the child can literally have a relation to the representation even though the representation does not relate in turn to a real-world referent. By contrast, natural selection is never directed at nonexistent states of affairs; it only makes a "choice" after the fact, between actually existing alternatives.[56] So natural selection does not exhibit intentionality, which is why there is no warrant for attributing representations to it. Dennett's apparent response is to insist on two levels of intentionality: one, natural selection has "reasons" without representations (Dennettionality); two, humans are "reason-representers" and "self-representers."[57] But the distinction is desperately confused. First, what we mean by a reason in the case of agents is a representation: to have a reason to do A is to represent A as something to be done. Second, and in consequence, our grounds for saying that natural selection has no representations are grounds for saying that it has no reasons and hence not even Dennettionality. I am sure that Fodor (and Granny)[58] will agree that we need intentionality reform: no intentionality without representation![59]

There is something disturbing too about Dennett's reasons for saying that biological function points to intentionality but not to intentional states. If natural selection only exhibits the illusion of foresight, surely the conclusion to draw is that it only exhibits the illusion of intentionality. If so, it is not true that it exhibits even a low-order intentionality or Dennettionality. In that case, Dennett's series of enhancements, from Darwinian to Gregorian creatures, cannot even get started. Further, if the reason that we should not infer representations is that natural selection is a blind mechanistic process, then Dennett lays himself open to the eliminative materialist who will argue that the blind mechanistic process of vector-to-vector transformation of neural-activation patterns accounts for cognition without intentional states. Of course, Dennett will say that "the real patterns" of human action are best captured by appeal to intentional states. But now that we have seen that the idea of reasons without representations is incoherent, nothing blocks the parallel argument that the best explanation of the functionality of biological structures is found in the intentional states of a nonhuman, and arguably nonmaterial, agent.[60] The dilemma Dennett faces is that if he denies agency as an explanation in biology, the eliminativist will push him to deny it in psychology, while if he affirms agency in psychology, the proponent of intelligent design will push him to affirm it in biology.

Objection 2: The Subjectivity Problem Revisited

Neither Dennett's account of the robotic survival machine nor his account of the transition from Darwinian to Gregorian creatures explains the emergence of agents capable of having personal reasons and subjective experiences.

We have already seen that standard physicalist accounts are plagued with an inability to account for the subjectivity of experience and the personal reasons of an agent. To be sure, Dennett does not claim that Mother Nature has experiences or personal reasons, but he does offer two pictures of how increments in complexity might lead to systems that possess them.

The first is the account of the robotic survival machine, designed so that it can derive its own goals from an assessment of its current state and its final goal. The problem with this proposal could not be more fundamental. Dennett gives no reason to think that the robot is an agent, that is, a coherent unity that is capable of pursuing its own goals. When agents have a goal, they represent themselves as having achieved it in the past or as achieving it in the future. But this ability is not explained by clever engineering, such as self-monitoring via feedback loops. Supposing that this produced a representation, it would simply be an impersonal representation of the system's current state. What is needed to have a self-concept is the concept of an agent extended over time so that it is one and the same agent who achieved certain goals and who aims to achieve others in the future. But before there can be the concept of an agent, there must first be an agent; and before Dennett's robot can have its "own goals," there needs to be something that can have goals. The system may contain states that the system's designers use to represent some goals of their own, but that does not mean the system is the kind of thing that can have its own goals. For as Quassim Cassam argues, "first-person thoughts are only correctly ascribable to persons."[61] (My only caveat with Cassam's statement would be to substitute "agents," in the robust sense defined in chapter 1, for "persons," in case there are subpersonal agencies, such as certain animals.) Likewise, to have subjective experiences, a subject must exist to have such experiences. A physical system may be able to detect pain or events associated with pain, but that is not the same as experiencing oneself as being in pain. This again requires a self-concept, which in turn requires a self or agent. Thus, Dennett's account of the robot's self-represetation simply assumes agency, without giving an account of how it emerges.

Dennett's second story tracks the transition from Darwinian to Gregorian creatures. He tells us that Skinnerian creatures could learn because they "confronted the environment by generating a variety of actions, which they tried out, one by one, until they found the one that worked."[62] Learning by experience is dangerous since a wrong move may be one's last. Popperian creatures are more advanced because each has "a sort of inner environment—an inner something-or-other that is structured in such a way that the surrogate actions it favors are more often than not the very actions the real world would also bless, if they were actually performed."[63] Without going any further, it is clear that Dennett is assuming that both Skinnerian and Popperian beings have a self-concept. Skinnerian beings learn that some past action of theirs is a guide to their future actions. If so, then they must think of themselves as agents extended over time. And since they learn from their own experience (from how things seem to them), they must be subjects of experience. Furthermore, what they learn is that some action is the one they should perform, which implies personal reasons. Likewise, Popperian beings use their inner environment to consider what would happen—and in some cases, what they would experience—if *they* performed certain actions. All of this presupposes the kind of self-representation that only agents possess. Thus, Dennett's account of the transition from Darwinian to more complex creatures does nothing to explain the emergence of agents capable of personal reasons and subjective experience, because they are already assumed.

Objection 3: The Sufficiency Problem

Dennett appears to equivocate between Minimalist Mother Nature, who is compatible with standard accounts of natural selection; and Hagiographical Mother Nature, who is not. Minimalist Mother Nature does not suffice for intentionality or even a significant enough ingredient to qualify as Dennettionality. Hagiographical Mother Nature does suffice for intentionality but only because she is endowed with powers of design incompatible with materialism.[64]

Searle points out that the location of intentionality in natural selection is highly unpromising because "intentional standards are inherently normative"; however, "there is nothing normative or teleological about Darwinian evolution."[65] At times, Dennett acknowledges this and describes a Minimalist Mother Nature. For example, in commenting on the idea of gradualistic "hill-climbing" in Darwinian evolution, Dennett says, "there cannot be any intelligent . . . foresight in the design process, but only ultimately stupid opportunistic exploitation of whatever lucky lifting happens your way."[66] This is Dawkins's picture of the blind watchmaker, and it surely excludes anything worth calling intentionality. When human beings select something, they select it because they think it will be best for them because of its anticipated but currently nonexistent good consequences. As we saw under the first objection, the ability to pursue "intentionally inexistent" states of affairs is essential to intentionality, but Minimalist Mother Nature never does this. As Fodor says, Minimalist "Mother Nature never rejects a trait because she can imagine a more desirable alternative, or ever selects for one because she can't. We do."[67] In the face of this objection, Dennett must either retreat to a fictionalist or instrumentalist account of Mother Nature's intentionality or admit that Mother Nature's intentionality is not as real as ours and show how it can be supplemented to get the real thing. If he takes the former option, the eliminativist is ready to force him into instrumentalism about human cognition as well. So what about the latter option?

Some have argued that a primitive form (or element) of intentionality arises naturally because Mother Nature does not merely select organisms or species but selects *for* certain traits.[68] The idea is that natural selection selects for those traits that actually have survival value so that the heart is selected for pumping blood and not making a rhythmic noise, even though all hearts make such a noise. As a result, "selected for *F*" is an intentional context since it may be that a structure was selected for *F* but not for G, even though all *F*'s are G's and even if *F* and G are coextensive. Consistent with Minimalist Mother Nature, the most plausible account of such intentionality is that the intentionality of "selects for" is inherited from the intentionality of "explains." That a heart pumps blood explains why organisms have a heart, but that a heart makes a rhythmic noise does not. Intuitively, the reason is that pumping blood is essential to survival, but making a noise is a mere by-product of pumping blood; additionally, had hearts appeared that pumped blood but made no such noise, they would still have been selected.

Supposing that such an account is correct, it does not get us very far. The statement "Hearts were selected for pumping blood" is consistent with Minimalist Mother Nature in that those organisms whose hearts pumped blood (well) survived long enough to reproduce, whereas those whose hearts did not pump blood (well) did not. This reasoning can be fully understood at the level of ordinary causation, with no recourse to even the beginnings of intentionality. Given two coextensive properties *F* and G, the fact that *F* but not G is causally relevant to some state of affairs is consistent with causation being a blind undirected connection, and for that very reason, causal relevance does not help explain how intentionality might be directed toward *F* and not to G.[69] This fact is obscured because "selected for" tends to smuggle in an illicit notion of teleology, suggesting that Mother Nature selected hearts so that they pump blood. This, however, is an unjustified attribution of purpose and foresight, incompatible with Minimalist Mother Nature.

Further, Minimalist Mother Nature is hampered because she has no representations. A human being can want a drug even though it is a natural law that anyone who has the drug will die and that a human being does not want to die. Consequently, in intentional contexts, there is failure of substitution for even nomologically coextensive terms. For that matter, a child might want four beads and not want the square root of sixteen beads, even though it is logically necessary that four is identical with the square root of sixteen. By contrast, "selected for" makes no such discriminations; as Fodor points out, "contexts of explanation are transparent to the substitution of (e.g., nomologically) necessary equivalents."[70] In a creature's habitat, if it is a law that all and only red toadstools are poisonous, then when the creature detects them by sight, their being red is just as good an explanation of the creature's avoiding them as their being poisonous.[71] And if a creature's having four stomachs is causally relevant to its survival, so is its having the square root of sixteen stomachs. Clearly, we do not get any significant start on intentionality without representation, so Dennett cannot avoid the need to develop a materialistic account of that. But as we have seen, at least on current showing, either such accounts do not work, or they invoke norms of proper functioning just as problematic for materialism as intentionality.

However, when Dennett wants to convince us that Mother Nature can account for intentionality after all, he falls back on the idea that Mother Nature is an artifactual designer. The concept of design is sufficient for the characteristics of intentionality. A drug may be designed to cure condition C and not cause side effect *E*, even though it is a law that anyone with C who takes the drug will develop *E*. A device may be designed to give four quarters, but not the square root of sixteen quarters, as change for a dollar. Indeed, a strategic defense system may be designed to protect against an attack that never occurs, so the notion of design even allows for "intentionally inexistent" states of affairs. The obvious problem is that attributing such powers of design to Mother Nature abandons the lean but respectable Minimalist Mother Nature in favor of the more voluptuous and seductive Hagiographical Mother Nature.

The kind of designing power attributed to Hagiographical Mother Nature presupposes that she has goals and foresight. If hearts are designed in order to pump blood, that means that Mother Nature selected them for that reason and consequently that Mother Nature represented pumping blood as an advantage of certain heart designs. If anything is literally designed, then a design must exist as a representation before the product of the design. At this point it becomes obvious that the designer must itself have intentional states, so it is not true that Dennett offers a reduction of human intentionality with representations to some putative nonrepresentational intentionality. If reduction is the goal, "You can't explain intentionality by appealing to the notion of design because the notion of design *presupposes* intentionality."[72] Hagiographical Mother Nature is not a crane for intentionality but a relocated skyhook. Any reductionist wanting a materialistic account for human intentionality would want exactly the same account of the intentionality of Hagiographical Mother Nature.

What makes matters worse is that, as a materialist, Dennett must deny that Mother Nature is an agent since this would be a retreat to "mind-first" explanations antithetical to materialism. But if Mother Nature designs creatures and is not an agent, Dennett is committed both to Mother Nature's being a designer and her not being a designer since only agents are designers. As Fodor says, "*That just makes no sense*."[73] What is more, the whole point of the theory of natural selection was to argue that nature is not designed but only exhibits the appearance of design. But if there is no design, then there is no Hagiographical Mother Nature: even if the idea of a nonagent's designing things were coherent, there is no work for her to do, "so it can't be that we are her children or her artifacts, or that our intentionality derives from hers."[74] The one remaining possibility, that Hagiographical Mother Nature is only a fiction, will be explored under the next objection.

In conclusion, Dennett offers Minimalist Mother Nature, who is insufficient to account for intentionality; or he offers Hagiographical Mother Nature, to whom is attributed the same kind of intentionality that needs explaining. And what is particularly absurd about the latter move is that only agents have this kind of intentionality; yet, as a materialist, Dennett must deny that Mother Nature is an agent.

Objection 4: The Incoherence Problem Revisited

Dennett's account of the intentional stance misses the primary reason it provides for believing in intentionality. The idea that using the stance requires us to think of ourselves as products of a fictional Hagiographical Mother Nature is unnecessary and incoherent.

I have agreed with Dennett and against Churchland that "the abstraction problem" provides sound evidence for the existence of intentional states. However, I think that Dennett is mistaken to think that this is our primary reason for believing in intentionality. It is not the predictive success of adopting the intentional stance but the fact that humans are able to adopt explanatory stances that is so telling, be they physical, design, or intentional stances. The irony is that the answer has been under our noses all the time. To adopt an explanatory stance toward something is to seek to understand its behavior by showing how it falls under certain kinds of concepts. Even adopting the physical stance requires us to understand a system in terms of its physical constitution and operation, but to do so is to view it in a certain way and to adopt an intentional attitude (understanding) toward it. Consequently, only beings who have particular points of view and intentionality can so much as adopt the physical stance, let alone the design and intentional stances. It is absurd to maintain that the various stances aim at understanding and then suggest that we need the empirical success of adopting the intentional stance as warrant for believing in intentional states. That we can adopt any of the stances shows that intentionality exists, and as a result, the intentional stance is predictively successful. Jennifer Hornsby is sensitive to this point, noting that Dennett overlooks the possibility that "those who use common-sense psychology to interpret (who 'take the Intentional Stance,' . . .) coincide with those who can be interpreted using common-sense psychology."[75] If I am right that adopting the intentional (or any other) stance requires intentionality, it is unsurprising that this should be so.

This observation also undercuts Dennett's claim that we need to view ourselves as products of Mother Nature to understand our intentionality. The very same intentionality that enables us to adopt the intentional stance toward others allows them to do the same to us. This also explains the obvious fact that folk psychology was successful long before anyone proposed the theory of natural selection or viewed him- or herself as its product. At a deeper level, if Hagiographical Mother Nature only appears to exist (she is at best a useful fiction),[76] then all she could explain is the appearance of intentionality. But in fact she could not even explain that, for if there is such a thing as the appearance of intentionality, then there must be real intentionality because "it appears that *p*" defines an intentional context. What needs explaining and what the fiction of Mother Nature cannot explain is precisely why there are such things as appearances.

Further, as Hornsby points out, if Mother Nature is fictional, it is worth asking how we construct the fiction.

> Presumably we model her on our (psychological) selves. But then we have to construct a fiction (Mother Nature) in order to tell a story . . . yet we have to construct the fiction out of material (common-sense psychology) that could only be available once the fiction had already been constructed and the story told.[77]

Once more the idea of constructing a fiction is thoroughly intentional since, by definition, fictions consist of intentionally inexistent states of affairs that the author understands and intends to communicate; so if we can construct fictions and understand that fact about ourselves, we have no need to think of ourselves as products of Mother Nature to understand our intentionality. What is more, if what is fictional about Mother Nature is her intentionality and if our intentionality derives from hers, then (in the absence of plausible cranes) it follows that our intentionality is fictional. But it is incoherent to claim that fictional intentionality is what enables us to construct a fiction of Mother Nature, since fictional entities do not exist—not even as abstracta. Dotheboys Hall is a fictional school described by Charles Dickens in *Nicholas Nickleby*. I think it is safe to assume that it is not Dotheboys Hall that explains why Dickens wrote a fiction about Dotheboys Hall.

Dennett cannot have it both ways. If the intentional stance shows that intentionality figures in "real patterns," then this cannot be explained by appeal to a fictional entity: This would only generate fictional patterns. Fairy stories about Mother Nature cannot be the explanation of the success of the intentional stance. On the contrary, it is the ability to adopt stances—and to construct fairy stories—that show that intentionality is real and that is the reason why the intentional stance works.

Science and the Rationality of Materialism

The upshot of our examination of orthodox physicalism and Dennett's significant alternative is that attempts to domesticate agency as a part of nature are faced with insurmountable obstacles. If the accounts remain faithful to materialist strictures, they are unable to account for the points of view, intentionality, and personal reasons of agents. Yet, if the accounts can explain these aspects of agency, then these very notions or others equally problematic for materialism have somewhere been smuggled in without justification. Appeal to cranes and incremental complexity does not work since there seems to be a fundamental difference of kind between blind mechanistic processes, no matter how complex the system, and the teleology and intentionality of thought. This is good reason to think that agency is not a "free lunch," something that simply emerges from materialist categories when they reach a certain level of complexity. Rather, the evidence is that Brentano was correct: intentionality (and agency in general) is sui generis and flatly inconsistent with materialism. Intentional patterns cannot be reduced to materialist categories.[78] If this is correct, then those like Dennett, who claim to be materialists but accept the reality of intentional patterns, hold an inconsistent position. The arguments for acknowledging intentionality and agency as fully real are reasons for abandoning materialism.

What, however, is the alternative to materialism? The answer becomes clear when we consider the real implications of our ideas about design, functions, and intentionality. As we saw, at one point, Fodor protests against Dennett that it is hopeless to account for our intentionality by appeal to the design of Mother Nature "because the notion of design *presupposes* intentionality."[79] This point suggests the following argument, reminiscent of Aquinas's Fifth Way:

P1: If something has a purpose, then it is designed.
P2: Intentionality has the purpose of guiding behavior.
P3: So intentionality is designed (via P1 and P2).
P4: But clearly, our intentionality is not designed by us, although it does enable us to convey our own designs.
P5: Thus, our intentionality is the result of prior design (via P3 and P4).
P6: But, as Fodor reminds us, if something is designed, then it is the product of intentionality.
C: So, our intentionality is the product of prior intentionality (via P5 and P6).

The argument naturally leads to the postulation of some agency prior to and arguably higher in power than any human agent, although it does not establish that this agency has all the characteristics ascribed to God by theism. I happen to think that this simple argument is correct and that the materialist has no compelling response. The materialists may claim that nothing, including intentionality, has a purpose. In that case, we do not act for purposes. So the materialists are advancing SAR and will find that they have undermined the rationality of science and any credible account of human's mutual understanding (see chapter 2). Or the materialists may say that human purpose is real but that it can be naturalized. So they are advancing WAR, which requires a materialistic account of intentionality. But as we have just seen, such accounts fail.

The remaining possibility for the materialist is to claim that my argument could not possibly be right because it invokes a "skyhook" and because appealing to a higher agency does not explain agency. It will be said that what I have done is no better than postulating Hagiographical Mother Nature since the problem of explaining human agency has simply been relocated to the problem of explaining some higher agency. Indeed, this explains why Dennett thinks that "mind-first" explanations are hopeless and nonexplanatory.[80] Likewise, Georges Rey claims that "any ultimate explanation of mental phenomena will have to be in non-mental terms, else it won't be an *explanation* of it."[81]

In response, I would first point out that not all explanation need be either ultimate or reductive. We can give a proximate explanation of the motion of billiard ball *B* by appeal to the prior motion of billiard ball A, even though this does not reduce motion to anything else. Likewise, if intentionality is real and does not reduce to anything else, we can still offer a proximate explanation of the origin of human intentionality in terms of some prior intentionality.[82] Indeed, we should because one of the many things Dennett is right about it is that we do not have "original intentionality."[83] This is surely highly implausible given our contingency. Not only our material states but also our intentional ones come into existence, although previously they did not exist; so they certainly cannot be self-explanatory. Nor are they explained as simply the effects of current physical processes in the brain since blind materialistic causes are insufficient to account for the teleology and foresight in the effects.[84] So if our intentionality is derivative, Haugeland is right to ask where this leads:

> Derivative intentionality, like an image in a photocopy, must derive eventually from something that is not similarly derivative; that is, at least some intentionality must be original (non-derivative). And clearly, then, this original intentionality is the real metaphysical problem; for the possibility of delegating content, once there is some to delegate, is surely less puzzling than how there can be any in the first place.[85]

The materialist assumes that this search for ultimate explanation must lead to a reduction of the intentional to the nonintentional. But my second point is that even a reductionist is wrong to exclude the possibility that intentionality is irreducible. For if explanation is ever to come to an end, it cannot be that every property is explained only by reducing it to something else. Indeed the goal of reduction is to identify some basic set of properties from which all the phenomena can be reconstructed. All other properties may be reducible to the basic properties, but the basic properties do not reduce to anything else. What, but a materialist bias, precludes the possibility that intentionality is one of these basic properties? Further, one cannot simply shrug off the demand for explanation as unnecessary, because human intentionality is contingent and hence neither self-explanatory nor a brute fact. As John Warwick Montgomery points out, to claim otherwise is

> to deny the contingent nature of [this feature of] the world and mythologically make it absolute—in the face of all empirical knowledge of its non-self-explanatory character.[86]

Third, traditional theism does have an explanation for the origin of human intentionality, one that shows that appeal to a higher agency need not be like appeal to Hagiographical Mother Nature. As traditional theism characterizes God, His agency requires no further explanation because the traditional God is an eternal and perfectly complete agent, a necessary being whose existence and attributes do not depend on the existence of anything else. Unlike us, the traditional God is not a finite, contingent agent but an infinite, necessary agent, and in that sense his agency is self-explanatory. Charles Taliaferro puts the point well:

> To court theism is to entertain the thesis that there is a being whose properties of omniscience, omnipotence, goodness and aseity are not derived from some other agency.[87]

The case for explaining contingent agency by appeal to a divine agent is particularly strong because, as Montgomery has emphasized, there is good reason to believe that "personality does not arise from the impersonal" and because human intentionality is clearly contingent. Thus, if human personality cannot arise from the impersonal matter of the universe, its source surely has to be a supernatural being; and on pain of regress, this being must be supposed to be a necessary or "Absolute" being: "the existence of personhood [in the world] is one of the contingencies requiring an appeal to a transcendent Absolute."[88]

At this point the materialist is liable to protest that God is the ultimate supernatural skyhook and hence cannot be invoked as a legitimate scientific explanation. In response, we may note that science has frequently benefited from the postulation of skyhooks, albeit typically natural ones (see chapter 1). But the real objection is not to nonmechanistic agency, which has proven fruitful in science, but to supernatural agency. The most frequent charge is that involving the supernatural would imply "gaps" in nature, which would make systematic scientific study impossible. As common as this argument is, it has been thoroughly discredited by the recent work of Del Ratzsch. He points out that scientists already have means for detecting when unaided nature could not produce a given phenomenon and for discerning what sort of agent was responsible. In archaeology, we may conclude that an item is a human artifact and not the result of a natural process. In the search for extraterrestrial intelligence (SETI), a signal exhibiting complex specified information would allow us to infer an alien intelligence. What if we found something that could not be explained by any contingent intelligence (human or alien), such as (if I am right) intentionality or agency itself?[89] Ratzsch spells out how a scientist who is not dogmatically wedded to materialism can argue:

> If unaided nature cannot generate some phenomenon, and there that phenomenon is in front of us, then obviously some other agency was involved. If we add the premise that humans couldn't or didn't produce the phenomenon, whereas aliens could have, we get the aliens-of-the-gaps arguments, which is precisely what underlies SETI. If we add the further premise that aliens couldn't or didn't . . . then supernatural agency follows.[90]

As we saw in chapter 1, science can make advances by showing that a reduction fails. The alchemical reduction of silver and gold to base metals, the Cartesian reduction of motion to direct contact, and the reduction of electromagnetic phenomena to the aether were all failures as reductions and yet huge advances for science. If contingent agency cannot be accounted for in materialistic terms yet can be explained by supernatural agency, then the bridge between science and theology, burned down by the Enlightenment's exclusion of the supernatural from science, stands rebuilt.

However, others allow that the idea of a supernatural agent is coherent, but they still deny that it can play a role in empirical science because the actions of any such being are likely to be inscrutable to human understanding.[91] The objection seems to conflate two ways in which actions can be inscrutable. First, the motives of an action can be inscrutable, yet the effects can be perfectly clear, for example, a school shooting; or, second, the effects can be inscrutable, for example, when an unscrupulous corporation makes a decision behind closed doors. Even if the motives of a supernatural action are inscrutable, it does not follow that its effects are. Indeed, the effects may be the same as ones observed in nature, with the exception that no natural causes were available to produce them. Thus, water is regularly turned into wine through a natural process. If, on a given occasion, wine appears spontaneously with no such process, the wine is no more inscrutable than natural wine, even if, like the wedding guests at Cana, we have no idea what the motive for making it was, who made it, or how it was made.[92] Some supernatural effects certainly may be inscrutable, either because our faculties are unable to discern them or because a supernatural agent acts in such a way that the effects are indiscernible from chance or law.[93] But there is no good reason to assume a priori that all such effects should be inscrutable. It may be that the supernatural agent wants to reveal its character to us.

Second, I am happy to acknowledge the fallibility and ignorance of human beings but for that very reason find the skeptic's argument unconvincing. As Chesterton once said, "We do not know enough about the unknown to know that it is unknowable."[94] In the case of the traditional God of theism, the idea that we know that God is unknowable claims to know too much: surely it is up to God, not us, to decide how much we can know of His actions. If so, it is an empirical question whether God has made His designing work accessible to us. Certainly, the emergence of modern science was made possible by individuals such as Johannes Kepler, who saw scientific investigation as a matter of thinking God's thoughts after Him.[95] The assumption was that despite our frail faculties, there is a natural affinity between human and divine reason and agency, which is one reading of the biblical claim that humans are made in God's image. Further, the very idea that science is possible depends on the assumption that the same order or logos at work in the cosmos is mirrored in human reason. The fact is that even materialist science requires the assumption that the world and the human mind are such that the mind can discover how the world works; it is just that the materialist seems unable to justify this assumption.[96]

The idea that we cannot explain human agency by appeal to a higher, supernatural agency thus turns out to rest on a prejudice. It is not that no such explanations are possible, since theism does not merely postpone the need for the very same kind of explanation that human agency requires. Nor is it the case that all appeal to the supernatural is necessarily unscientific since there are scientific procedures for determining what unaided nature cannot do, and these can be extended to show the limits of contingent agents as well. And nor are there convincing a priori arguments to show that the effects of supernatural agency are necessarily inscrutable to human reason. Rather, the real reason so many reject theistic explanations is simply that the explanations are not materialistic or reductive, which simply begs the question. If materialistic accounts of agency cannot explain it but at least one nonmaterialistic explanation can, those with an open mind will follow the better argument. By acknowledging the reality of agency, the proponent of WAR will find that an unbiased commitment to the rationality of science requires the rejection of materialism as an a priori doctrine.[97]

Conclusion

In the last chapter, we saw that if the scientific materialists pursue SAR and attempt to eliminate agency, then they cannot sustain human practical rationality and hence cannot account for the rationality of science. They therefore lose the right to call themselves scientific materialists. In this chapter we have examined WAR, which claims that human agency exists but as part of the material world. This program has been shown to fail because there is no satisfactory, authentically materialist account of the points of view, intentionality, and personal reasons of agents. Indeed, human agency is best explained by a higher, supernatural agency. Thus, if the proponents of WAR retain an unbiased commitment to scientific evidence and accept the reality of human agency, they then lose the right to call themselves scientific materialists. Either way, scientific materialism is false.

Notes

1. Daniel Dennett, "Evolution, Error and Intentionality," in his *The Intentional Stance*, 298–99.

2. Jerry Fodor, "Deconstructing Dennett's Darwin," in *In Critical Condition: Polemical Essays on Cognitive Science and the Philosophy of Mind* (Cambridge, MA: MIT Press, 2000), 186.

3. It is notoriously difficult to say what the extension of "physical theories" is without making physicalism either trivially true or obviously false. As Crane and Mellor point out, if the term denotes current physical theories, then all sorts of new entities that may be discovered by future science would wrongly be excluded from the physical; so physicalism is obviously false. However, if the term denotes our best future physical theories (or even an ideal completed physics), then a priori, there is no way to exclude the possibility that these theories will refer to intentional mental states; and if these are physical, physicalism is trivial. The fact is that we cannot tell in advance whether some version of psychology might count as a physical science. Yet physicalism gets its interest from the assumption that there are some problematic entities, such as intentional states, that must either be reduced to unproblematic physical entities or be eliminated. See Crane and Mellor, "There Is No Question of Physicalism," *Mind* 99, no. 394 (April 1990): 188.

4. See Jaegwon Kim, *Mind in a Physical World: An Essay on the Mind-Body Problem and Mental Causation* (Cambridge, MA: MIT Press, 1998), ch. 4.

5. Kim, *Mind in a Physical World*, 106–12.

6. Gilbert Ryle, *The Concept of Mind* (London: Hutchinson, 1949).

7. While Ryle's point is well taken, there is a parallel, meaningful question about why all and only these colleges belong to Oxford University. That question cannot be answered by listing the colleges, and indeed it would be a category mistake to say that the list explained what was on the list.

8. In Searle's usage, Strong Artificial Intelligence claims that the appropriately programmed computer actually has understanding and intelligence, while Weak Artificial Intelligence claims that such a computer only simulates understanding and intelligence.

9. Searle, *The Rediscovery of the Mind*, 43–44.

10. One can debate this point because the program's functionality is dependent on the intentions of its designers, and if the function cannot be specified independently of those intentions, it begs the question in favor of physicalism to assume that the function is physical. However, a program can supply a physical notion of a function if it is understood minimalistically as a set of mappings between formally specified inputs and internal states and between such states and output. This does not beg any questions since the transitions are blind, automatic, and mechanistic, as the physicalist requires.

11. Ned Block pointed out that a computer program for pain behavior might be implemented manually by the entire population of China, without its being true that the population as a collective entity has any pain. See his monumental "Troubles with Functionalism," in *Minnesota Studies in the Philosophy of Science*, vol. 9 (Minneapolis: University of Minnesota Press 1978), 261–325.

12. In cases of inverted qualia, the causal mappings between input, internal states and output would be the same, but what would look red to one person would look green to another. This difference of appearance is consistent with identical physical internal states and behavior. The classic statement is Block and Fodor's "What Psychological States Are Not," *Philosophical Review* 81 (1972): 159–81.

13. Kim agrees that the absent–inverted qualia arguments show that qualia have no obvious functionalization. See Kim, *Mind in a Physical World*, 101–3.

14. Some philosophers hold that there can be natural representations independent of minds. They claim, for example, that, independent of human minds, tree rings represent the age of a tree because of the way they have been generated. A classic defense of this view is found in Dennis Stampe's "Towards a Causal Theory of Linguistic Representation," in *Contemporary Perspectives in the Philosophy of Language*, ed. Peter A. French, *Midwest Studies in Philosophy*, vol. 2 (Morris: University of Minnesota, 1977), 81–102.

15. John Searle, "Minds, Brains and Programs," *Behavioral and Brain Sciences* 3 (1980): 417–57.

16. In fact, the physicalist is not even entitled to the notion of syntax. Since all syntactic items reduce to binary representations and these in turn to the state of the computer's switches (on or off), what the computer does can be fully accounted for in terms of the configuration of its switches. Any particular syntax is merely an interpretation of these configurations. Searle makes a similar point in his later work: "For the purposes of the original argument, I was simply assuming that the syntactical characterizations of the computer was unproblematic. But that is a mistake. There is no way you could discover that something is intrinsically a digital computer because the characterization of it as a digital computer is always relative to an observer who assigns a syntactical interpretation to the purely physical features of the system" (Searle, *The Rediscovery of Mind*, 210).

17. A common misunderstanding is to claim that the "right kind" of symbol manipulations—for example, ones using sophisticated heuristics, feedback loops, multi-

layered neural connections, and so forth—are the key to understanding and intelligence. This may provide impressive simulations of understanding and intelligence, but no matter how sophisticated the transformations may be, so long as they are performed on patterns independent of their meaning, there is no reason to think that representation or understanding is doing any work. Of course, the output may be highly intelligent in that it would take a human a great deal of intelligence to produce it, but this merely shows that difficult but well-defined tasks can be automated—for example, by implementing an algorithm that maps a formal problem structure to a formal solution structure.

18. Consequently, as Searle says, the symbols have "derived intentionality"; that is, although they really can convey intentional content, this does not reside in the symbols themselves but derives from the intentions of language users. Only the users have "intrinsic" intentionality, intentionality that resides in them. As for the computer itself, Searle would say (and I agree) that it does not have any representations, so all we can say is that it is as if it had intentionality. See Searle, *The Rediscovery of the Mind*, 78–82.

19. Thus the programmer may say, "This is a program to find the Greatest Common Divisor" and write instructions to implement the algorithm in a programming language like C++. The meaning of these instructions, however, is merely the derived intentionality of language, not the intrinsic intentionality of mental states. Further, since these instructions are represented in the programmer but not the computer, they are not reason to say that the computer has even derived intentionality.

20. Searle, *The Rediscovery of the Mind*, 51.

21. See, for example, Ruth Millikan, *Language, Thought and Other Biological Categories: New Foundations for Realism* (Cambridge, MA: MIT Press, 1984).

22. "It appears that *p*" defines an intentional context. For example, it can appear that Santa is in the store, even though there is no Santa. To say that there only appears to be intentionality is self-refuting, as I note in chapter 2. This is a kind of ontological argument for intentionality. From the fact that there appears to be any it follows that there is.

23. A seminal work is Dennis Stampe's "Towards a Causal Theory of Linguistic Representation" (see n. 14).

24. Jerry Fodor, *Psychosemantics: The Problem of Meaning in the Philosophy of Mind* (Cambridge, MA: MIT Press, 1987).

25. Causal theories often invoke the idea of what normally causes a given state or what causes it under optimal or ideal conditions. To the extent they do, they are not materialistic theories since materialistic causation is simply a blind automatic connection between events with no purpose in mind. That flicking a light switch ought to turn the light on is not derivable from the materialistic causal connections between the two events, but it reflects the purposes of human electricians in designing and installing the circuit.

26. For example, a strategic defense system may be supposed to protect against an attack that never occurs and that is consequently an "intentionally inexistent" state of affairs.

27. I have heard it said that this is a pseudoproblem, that the reason a given set of representations belong uniquely to one agent is that they all occur in him or her. This does not work, because my representations remain atomistic and unconnected from one another; unless, however, there is some common principle (Kant called it the "transcendental unity of apperception") in virtue of which the agent can think that these are all his or her representations. The unity of an agent's representations cannot be understood after the model of a bucket of seashells. How does the bucket (the mind) think of all the seashells (its thoughts) as its own? That is the question.

28. This is precisely the danger Crane and Mellor warned about in their article, "There Is No Question of Physicalism."

29. This irony is at its greatest in one of Dennett's replies to Thomas Nagel. Nagel had argued that materialism is inadequate because it makes no room for irreducible points of view. Dennett's reply is that the only way to find out if there are really limitations to materialism is "to see . . . just what the mind looks like from the third-person, materialistic perspective of contemporary science." (See Dennett's "Setting Off on the Right Foot," in his *The Intentional Stance*, 7.) This model, however, only seems to fit Dennett's physical stance and perhaps his design stance. Dennett's intentional stance warrants the attribution of intentional states that are an agent's personal reasons for action. This perspective is not obviously compatible with "the third-person, materialistic perspective of contemporary science."

30. Dennett, "True Believers: The Intentional Strategy and Why It Works," *The Intentional Stance*, 16.

31. Dennett, "True Believers," *The Intentional Stance*, 16–17.

32. Typically we assume optimality ("innocence") until we find a design flaw ("proven guilty"), at which point we downwardly revise to the nearest approximation to optimality consistent with the design flaw.

33. Dennett, "True Believers," *The Intentional Stance*, 17.

34. Dennett, "Three Kinds of Intentional Psychology," *The Intentional Stance*, 53.

35. Christopher Viger, "Where Do Dennett's Stances Stand?" in *Dennett's Philosophy: A Comprehensive Assessment*, ed. Don Ross, Andrew Brook, and David Thompson (Cambridge, MA: MIT Press, 2000), 134

36. Dennett, "True Believers," *The Intentional Stance*, 29.

37. Christopher Viger, "Where Do Dennett's Stances Stand?" 137.

38. See Dennett, "Evolution, Error, and Intentionality," *The Intentional Stance*, 287–321.

39. The example is Searle's. See *The Rediscovery of Mind*, 78. Searle, however, seems less comfortable with the idea of derived intentionality beyond language.

40. Dennett, "Evolution, Error, and Intentionality," *The Intentional Stance*, 297.

41. Richard Dawkins, *The Selfish Gene* (Oxford: Oxford University Press, 1976).

42. Likewise, Dawkins frequently makes the point that via our memes, we can transcend the dictates of our genes.

43. Dennett, "Evolution, Error, and Intentionality," *The Intentional Stance*, 299.

44. Dennett, "Evolution, Error, and Intentionality," *The Intentional Stance*, 305.

45. Dennett, "Evolution, Error, and Intentionality," *The Intentional Stance*, 315. We need the intentional stance to do reverse engineering since we must assume that a biological structure is for some work to figure out how it does that work.

46. Dennett, "Evolution, Error, and Intentionality," *The Intentional Stance*, 316–17.

47. Ruth Millikan argues that we can understand human thought in terms of the evolutionary "proper functioning" of our cognitive mechanisms. See her *Language, Thought and Other Biological Categories* (Cambridge, MA: MIT Press, 1984).

48. Dennett, "Evolution, Error, and Intentionality," *The Intentional Stance*, 300.

49. Dennett, "Evolution, Error, and Intentionality," *The Intentional Stance*, 318.

50. As a theist, I see human intentionality as ultimately deriving from God's. Indeterminacy strikes me as an obvious consequence of the fact that people fail to gain mastery of certain concepts. I deny the Burgean thesis that we should attribute intentional states about arthritis on the basis of the correct public use of the linguistic term "arthritis." (See Tyler Burge's "Individualism and the Mental," *Midwest Studies in Philosophy* 4: 73–121; and his "Individualism and Psychology," *The Philosophical Review* 95, no. 1: 3–46.) If someone really cannot distinguish arthritis from other similar ailments, then his statement that he has arthritis may be unproblematically false because statements are held accountable to public standards of linguistic usage. But I do not think it obviously true that we should ascribe to this person the belief that he has arthritis in his joints (or even that he has arthritis or similar ailments) because his concept of "arthritis" is not determinate enough to make these discriminations. The vagaries of what, if anything, natural selection "selects for" are irrelevant since the problem of indeterminacy arises independently from a consideration of conceptual mastery. Furthermore, it is a red herring in the current debate since our difficulty in expressing certain intentional contents in public language has nothing to do with whether they really exist. It merely shows that the standards to which we hold language users accountable are sometimes too high to capture people's mental contents accurately. Hence I am sympathetic to the idea that folk psychology can be reformed by the addition of "narrow content," a kind of content that tracks the way the world looks from a subject's conceptual point of view more closely than "broad content," which links the subject to her public environment.

51. Dennett, "Evolution, Error, and Intentionality," *The Intentional Stance*, 317.

52. See chapter 1 of this book and Dennett, *Darwin's Dangerous Idea*, 76.

53. Dennett, *Darwin's Dangerous Idea*, 82–83.

54. Dennett, *Darwin's Dangerous Idea*, 374–78.

55. Fodor argues at length that evolution does not explain intentionality, because it cannot account for mental relations to what Brentano called "intentionally inexistent" objects, such as Santa Claus. See Fodor, "Deconstructing Dennett's Darwin," in his *In Critical Condition* (Cambridge, MA: MIT Press, 2000), 182–84.

56. The long and the short of it is that unfit creatures do not live long enough to reproduce. But it is not true that this was something natural selection intended, since intentions are directed at the future and natural selection has no prevision (or, indeed, any other kind of vision).

57. Dennett, "Evolution, Error, and Intentionality," *The Intentional Stance*, 317.

58. "Granny" is invoked as a voice of common sense in many of Fodor's writings.

59. A much better response for Dennett to make would be to claim that Dennettionality is not yet intentionality at all but that it does contain a crucial ingredient of it. Then he needs to explain in detail how the progression from Darwinian to Gregorian creatures uses cranes to supplement and magnify that ingredient. This gradualistic approach is considered later.

60. Nothing, that is, except Dennett's antipathy to "mind-first" explanations, to which we will return in the next section.

61. Quassim Cassam, "Reductionism and First-Person Thinking," chapter 13 in *Reduction, Explanation, and Realism*, ed. David Charles and Kathleen Lennon (New York: Oxford University Press, 1992), 361–80, 369. Cassam gives powerful reasons for rejecting Parfit's thesis that personal identity can be reduced to impersonal facts of psychological continuity.

62. Dennett, *Darwin's Dangerous Idea*, 374. Skinner, to be sure, would not accept Dennett's mentalistic description of operant conditioning. Skinner saw conditioning as the physical development of stimulus–response connections, in which representations played no role.

63. Dennett, *Darwin's Dangerous Idea*, 375.

64. A quite different line to take against Dennett is to question his adaptationism. Fodor has pointed out that even if neurophysiological structures were selected for, it does not follow that this was because of the associated psychological capacities. See Jerry Fodor, *The Mind Doesn't Work That Way* (Cambridge, MA: MIT Press, 2000), especially 87–90. I return to this issue in chapter 5, where I critique evolutionary psychology.

65. Searle, *The Rediscovery of the Mind*, 51.

66. Dennett, *Darwin's Dangerous Idea*, 191.

67. Fodor, "Deconstructing Dennett's Darwin," 182.

68. Elliott Sober defines "selection for" in his *The Nature of Selection* (Cambridge, MA: MIT Press, 1984), 97–102. Sober himself does not make any grandiose connections between this idea and intentionality.

69. This criticism is directed at standard Humean accounts of causation that are void of teleology. It has no force against the richer notion of causation recently defended by Robert Koons in his *Realism Regained: An Exact Theory of Causation, Teleology, and the Mind* (New York: Oxford University Press, 2000). However, Koons's notion of causation is certainly incompatible with materialism and hence no comfort to the defender of WAR.

70. Fodor, "Deconstructing Dennett's Darwin," 180.

71. Perhaps a better one because if the creature is a Skinnerian creature (who occasionally makes the mistake of trying a nonlethal amount of the poisonous toadstools), we can predict that if its habitat changed to include nonpoisonous red toadstools, these would be avoided, too, because of the association with poison.

72. Fodor, "Deconstructing Dennett's Darwin," 177.

73. Fodor, "Deconstructing Dennett's Darwin," 178.

74. Fodor, "Deconstructing Dennett's Darwin," 186.

75. Jennifer Hornsby, "Physics, Biology, and Common-Sense Psychology," in *Reduction, Explanation, and Realism*, ed. David Charles and Kathleen Lennon (New York: Oxford University Press, 1992), 155–77, 175.

76. The fictional reading of Mother Nature as natural selection viewed from the intentional stance is supported by the fact that Dennett says that we are "more literally" the result of natural selection than Mother Nature (see the first quote at the head of the chapter). Elsewhere, Dennett says that intentional states are "attributed in statements that are *true* only if we exempt them from a certain familiar standard of literality," and he makes it clear that he does not endorse fictionalism in general (see Dennett's "Instrumentalism Reconsidered" in his *The Intentional Stance*, 72). By "non-literal," Dennett means that intentional states are abstracta, not concreta. However, as we have seen, there is solid reason to think that Mother Nature is an outright fiction.

77. Jennifer Hornsby, "Physics, Biology, and Common-Sense Psychology," 168.

78. In support of this thesis, George Bealer argues that intentionality "entails the existence of an *objective level of organization* that is non-physical." See George Bealer, "Materialism and the Logical Structure of Intentionality," in *Objections to Physicalism*, ed. Howard Robinson (Oxford, University Press, 1993), 101–26, 101.

79. Fodor, "Deconstructing Dennett's Darwin," 177.

80. Dennett makes this case in the first three chapters of his *Darwin's Dangerous Idea*.

81. Georges Rey, *Contemporary Philosophy of Mind* (Oxford: Blackwell, 1997), 21.

82. Furthermore, with an enriched notion of causation—for example, the idea of top-down emergent causation on which Koons and others are currently working—one may also be able to account for the causation of intentional states in the natural world. However, I am not here offering any account of mental causation or the mind–body problem; my focus is solely on the ultimate origin of intentionality.

83. We do, of course, have our own goals. To say that our intentionality derives from a higher being's does not imply that our goals are the same as the higher being's, only that we would not have any intentional states unless such a being existed.

84. If, like Searle, one claims that it is simply a fact about brains that they have the power to generate intentional states, one needs to identify the feature that brains have (but computers lack) in virtue of which brains (but not computers) have this power.

85. John Haugeland, "The Intentionality All-Stars," *Philosophical Perspectives* 4, *Action Theory and the Philosophy of Mind* (1990): 383–427.

86. John Warwick Montgomery, *Tractatus Logico-Theologicus* (Bonn, Germany: Verlag für Kultur und Wissenschaft, 2002), 3.8521, 118.

87. Charles Taliaferro, "Naturalism and the Mind," in *Naturalism: A Critical Analysis*, ed. William Lane Craig and J. P. Moreland (London: Routledge, 2000), 133–55, 151. ("Aseity" means self-existence.) Taliaferro also points out that, in a sense, the theist is a reductionist: while the materialist thinks that agency reduces to matter, the theist thinks that matter reduces to God's agency.

88. John Warwick Montgomery, *Tractatus Logico-Theologicus*, 3.8542, 118.

89. It doesn't work to suggest that our intentionality might derive from that of superior aliens, because they also are contingent beings; so the origin of their intentionality is left unexplained as well. That is why appeal to an eternal and necessary agent, whose intentionality does not derive from anywhere else, is explanatory.

90. Del Ratzsch, *Nature, Design and Science: The Status of Design in Natural Science* (Albany, NY: SUNY Press, 2001), 119. Ratzsch considers a variety of other reasons for excluding the supernatural from science and finds them all inadequate.

91. Some of these issues are handled in more depth later in the book.

92. I am not suggesting by this example that all phenomena that warrant a supernatural explanation are direct miracles. While I think intentionality has an ultimately supernatural explanation, I take seriously the suggestion that it could be built into nature, but such an enriched notion of nature is certainly incompatible with any nontrivial version of materialism.

93. Thus Dembski's filter will not detect the design of an agent who simulates chance or law.

94. G. K. Chesterton, *The Quotable Chesterton*, ed. G. J. Marlin, R. P. Rabatin, and J. L. Swan (Garden City, NY: Image, 1987), 386, quoted in William Dembski's *Intelligent Design: The Bridge between Science and Theology*, 60.

95. For more on this, see my "Thinking God's Thoughts after Him: How the Bible and Science View the World," editorial in *Issues in Christian Education* 35, no. 1 (Spring 2001): 4–5.

96. It may be said that evolution itself explains the affinity of man's mind to his natural environment. In chapter 6, I argue that this is not the case. A good reason to doubt the materialist's claim is given by Robert Koons, who persuasively argues that the authority of science to tell us what is in our ontology depends on an assumption of epistemic reliability that cannot be justified on materialist assumptions. That the laws of nature are such that the human mind can discover them cannot be explained by the laws themselves (including natural selection) but requires a supernatural explanation. See Robert Koons, "The Incompatibility of Naturalism and Scientific Realism," *Naturalism: A Critical Analysis*, ed. William Lane Craig and J. P. Moreland (London: Routledge, 2000), 49–63.

97. This is a vindication of Phillip Johnson's "wedge" argument, which aims to show that empirical science is not the same as materialistic science and that empirical science is better off without materialism.

CHAPTER FOUR

Bait and Switch: Indirectness and Biological Unity

> Nature does not have the foresight to put together a sequence of mutations which, for all they may entail temporary disadvantage, set a lineage on the road to ultimate global superiority.[1]

> No sound functional analysis is complete until it has confirmed (as much as these points ever can be confirmed) that a building path has been specified.[2]

> Ultimately there is, of course, absolutely no reason why functional organic systems should form the continuum that evolution by natural selection demands. In the world of physics and chemistry many phenomena are discontinuous. One cannot gradually convert one molecular species into another, neither can one convert gradually one type of atom into another. Between such entities there are jumps. Might not functional organic systems be similarly separated by discontinuities?[3]

In the last chapter, we examined weak agent reductionsim (WAR's) attempt to naturalize intentionality via the notion of function. It was argued that if functions are defined in materialistic terms, then they are unable to account for intentionality. This leaves the proponent of WAR three main options. The first is to attempt a materialistic explanation of intentionality in nonfunctional terms. This seems unpromising since intentional states are governed by norms of rationality, and these norms appear to specify the function of the mechanisms that produce these states. For example, it is a norm that if someone believes a conditional and its antecedent, then that person should believe the consequent, that is, barring independent evidence at least as strong to disbelieve it. But given such norms, it is hard to deny that a function of an agent's belief-forming mechanism is to provide one with beliefs supported by evidence. A second option for the proponent of WAR is to enrich the concept of nature so that it makes room for teleological causation. This accommodates a robust notion of function since teleological causation can select something for a purpose, which in turn allows an account of intentionality in terms of the proper function of our cognitive systems.[4] But this means rejecting genuine materialism. No materialist worth his salt will concede the existence of teleological causation. Mindless matter cannot have goals of its own, so if it does implement goals, they must be those of some prior mind. Teleological causation thus implicates the very kind of "mind-first" account that the materialist rejects.[5] If the proponent of WAR insists on retaining materialism, then the remaining option is to reject intentionality along with robust notions of function. This is a retreat to SAR, which, as I argue in chapter 2, is incompatible with the rationality of science.

Let us bracket that objection, however, and consider whether the materialist can give an adequate account of biological function. Both SAR and WAR claim that our natural environment contains the appearance of design and functionality. Even the most skeptical form of SAR, one that denies any literal form of adaptionism,[6] concedes that many biological structures appear to subserve functions. The materialist can avoid commitment to teleology by a debunking approach to biological function, arguing that robustly teleological functional terms are merely a useful fiction for simplifying the analysis of complex systems. In this way the materialist can justify the Darwinist[7] tactic of bait and switch, replacing the problem of explaining biological function, in the robust sense, with that of explaining the appearance of such function. Even so, the materialist must grant that biological systems exhibit a complex organization, coherence, and unity that distinguish them from typical, nonliving structures studied by chemistry and physics. It is the burden of this chapter to show that in at least some cases the materialist cannot account for these characteristics, because it cannot explain the unification of diverse parts coadapted to perform a common function. The attempted bait and switch fails because materialism is unable to explain even the appearance of biological function.

This undercuts SAR and WAR at the very foundation. SAR and WAR both assume that their real work is to eliminate or naturalize psychological categories because biological categories are unproblematic for materialism. In fact, biology is thought by many reductionists to be the right place to look for a materialistic explanation of psychology. This project falls to the ground if biological functions cannot be explained in materialistic terms. There is no use in the materialist's eliminating or naturalizing psychology in favor of neurophysiology if the correct explanation of the latter invokes an immaterial intelligent designer. The materialist cannot benefit from Millikan's attempt to show that thought is a functional category of biology if materialism cannot even account for the appearance of function.[8]

The clearest class of biological systems that resist materialist explanation are those that Michael Behe has termed "irreducibly complex." In this chapter, I begin by analyzing and illustrating this notion. Next, I consider the main lines of reply to Behe. Some critics acknowledge the existence of irreducibly complex systems but argue that they could still have formed gradually; other critics deny that irreducible complexity really exists; yet others admit that irreducibly complex structures could not form alone but suggest that they may benefit from an additional support system. It will be shown that none of these ingenious ideas works. This, I further argue, is because of a quite general weakness of Darwinian gradualism. Undirected processes cannot solve the problems of coordination and interface compatibility that arise for finely tuned, multipart systems. These problems are not peculiar to biology but arise for complex systems in other fields, such as computer science. The consistent materialist must assume that complex structures are built bottom-up by atomistic elements that cannot communicate with one another or be guided by common goals. This problem-solving strategy cannot account for the mutual compatibility and coordination required for diverse parts to subserve a unified function. Rather, this requires top-down design. But in a top-down design, a representation of the form of the final solution precedes the discovery of the details. This, however, implies teleology, which is inconsistent with both Darwinism and any strict version of materialism.

Irreducible Complexity

Although it may have some loopholes, let us begin with Behe's original definition of irreducible complexity.[9]

> (IC) An irreducibly complex system is "a single system composed of several well-matched, interacting parts that contribute to the basic function, wherein the removal of any one of the parts causes the system to effectively cease functioning."[10]

As an example outside biology, Behe considers a standard mousetrap (figure 4.1). This has five "well-matched, interacting parts," namely a wooden platform, a metal hammer, a spring, a catch, and a metal bar that holds back the hammer. If any one of these parts is removed, the mousetrap no longer traps mice. According to Behe, this means that an undirected, gradual process such as natural selection could not produce such a mousetrap. Natural selection can only build complex systems by starting with simpler systems and gradually making small improvements. To do this, however, it is necessary that at each stage there is a system that performs some selectable function. But none of the mousetrap's precursors are functional mousetraps. It follows that if these precursors have no other selectable function or if trapping mice is a function on which the "survival" of the system depends, then the mousetrap has no functional precursor consisting solely of a proper subset of the mousetrap's parts. Under these rather strong assumptions (too strong, as we shall see), natural selection could not generate a mousetrap.

Part of the reason that a mousetrap's primary function is so easily disrupted is that it exhibits nonredundancy. A system is nonredundant if it contains no duplicate or near-duplicate parts that could serve as surrogates for parts currently in use were the latter removed. Suppose the mousetrap's hammer is removed. Without extensive modification, no other part can play the role currently played by the hammer. Furthermore, without modification of the mousetrap's design, none of the other parts is available to be taken from its current role to serve in a different capacity. Using the holding bar as the hammer would mean both modifying the holding bar and removing it from

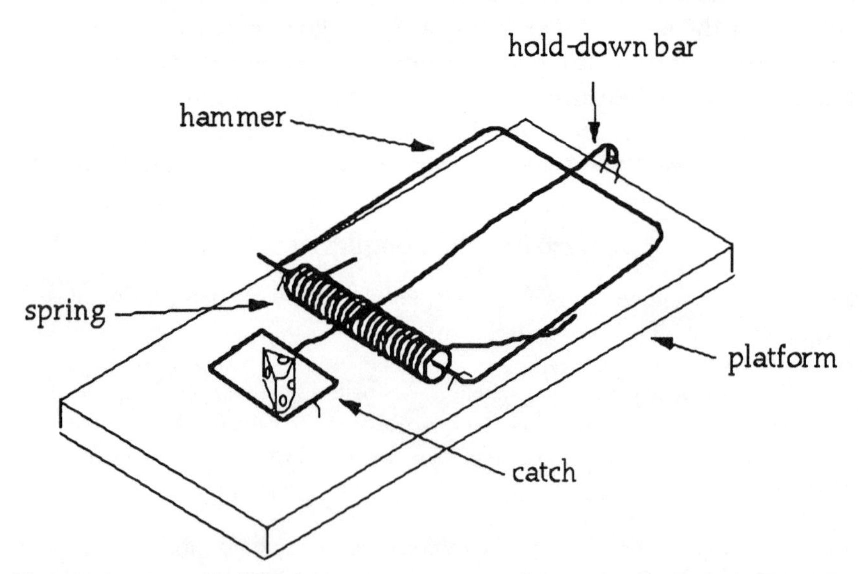

Figure 4.1. A standard five-part mousetrap. (Copyright © 2001 John H. McDonald. Reprinted with permission. All rights reserved.)

service as a holding bar. The holding bar cannot both be the hammer and what holds back the hammer.[11] But both are essential to the mousetrap's function. Using the holding bar as a hammer would be robbing Peter to pay Paul: the mousetrap would regain one essential component only by losing another. Function would therefore not be restored.

Nonredundancy is important because it helps to distinguish irreducibly complex systems from systems that are reducibly complex. *Reducibly complex systems* are ones that can continue to perform their primary function when a part is removed.[12] Consider the technique of double-stitching clothes. Double-stitching is a redundant system for holding garments together. If it is disrupted by the removal of a single line of stitching, the same function is still performed, albeit less reliably. Double-stitching is thus reducibly complex because it can be reduced to the simpler system of single-stitching and yet retain the same function of holding garments together. Thus it is entirely plausible that a process such as natural selection could select single-stitching and later select the more reliable system of double-stitching.[13] In general, redundancy means that the capacity of a system to implement a function is retained even if some parts of the system are removed. It follows that any system that is irreducibly complex in Behe's sense will be nonredundant.[14]

So much for the basic definition. Are there irreducibly complex structures in biology? According to Behe, there are many,[15] but we will focus on one of the clearest, the bacterial flagellum (figure 4.2).[16] The flagellum is a swimming device used by some bacteria. It includes a propeller (the filament) and a rotary motor. A rotor, the "M ring," is free to rotate, and it spins the filament; a stator, or "S ring," is a stationary ring mounted in the cell wall. Bushings, a drive shaft, and a hook joint, which connects the filament to the drive shaft, are also required. The motor draws its energy from the flow of acid found in the bacterial membrane. Essentially, the flagellum is an outboard motor for bacteria. It has quite astonishing performance, spinning at nearly 20,000 rpm yet able to reverse direction in a quarter turn.[17] It turns out that such performance is required for the bacterium to overcome Brownian motion and stay on track when following a glucose gradient.[18]

Even at an abstract level of description, it is clear that the flagellum is irreducibly complex. As Behe points out, all effective swimming devices must have something that plays the role of a paddle, to push against the water; a rotor, to repeat the action; and a motor, to supply the power. It is clear that the flagellum has all three of these parts, that they are "well-matched" and "interacting," and that the flagellum cannot perform its basic function if any one of these parts is removed.[19] But this design implies no simpler precursor of the flagellum that performs the same function using a proper subset of its

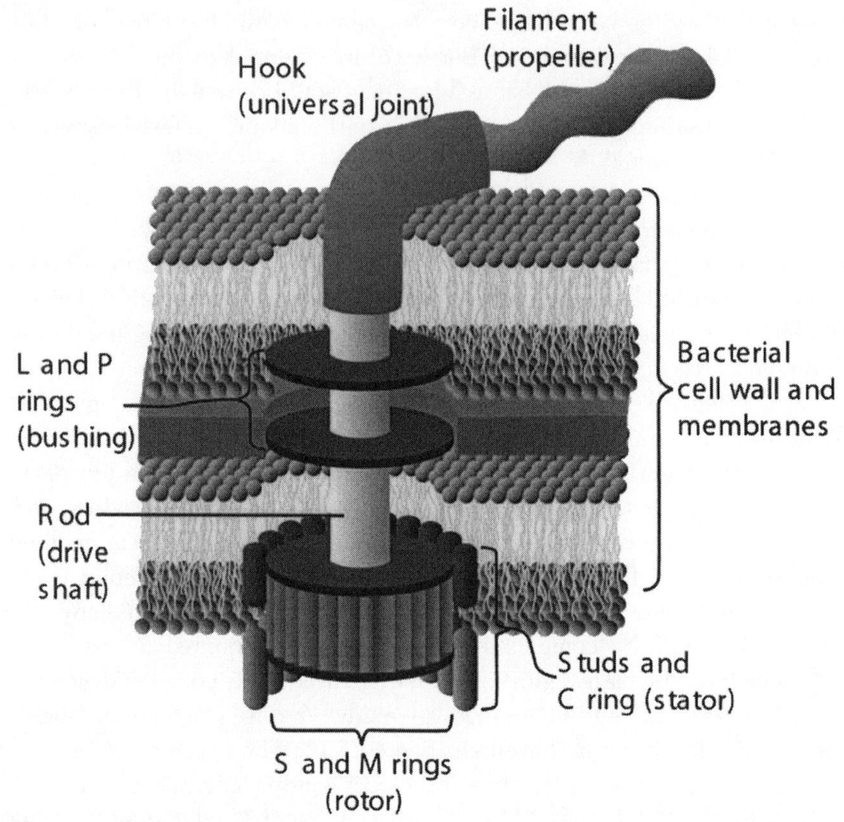

Figure 4.2. The bacterial flagellum. (Copyright © 2002 Timothy G. Standish. Reprinted with permission. All rights reserved.)

parts. It follows that a direct, gradual Darwinian path could not have built the flagellum from its parts by continuously improving its basic function.[20]

The challenge to Darwinism is particularly strong because the major parts of the flagellum are so finely tuned to its primary function that it is implausible that they could serve some other function on their own. Nor could natural selection select these parts because of their future function in the assembled flagellum. Natural selection has no foresight and therefore cannot select parts because they will be of use to a bacterium. Natural selection can only select a part because of its current use. But of what use is a rotor without a paddle and a motor? Of what use is paddle without a rotor and a motor? And of what use is a motor without a paddle and a rotor? The usefulness of any one of the parts is parasitic on the availability of the others. But then it seems that natural selection must build all of the parts at once and be able to coordinate them and ensure their mutual compatibility—and this is just too improbable to be credible. If preadaptation[21] on this scale occurs, it cannot be maintained that evolution is undirected.

The challenge is even starker when we move to a more detailed level of description. The flagellum's motor alone "requires the coordinated interaction of about thirty proteins and another twenty or so proteins to assist in their assembly."[22] If it turns out that some of these proteins are also individually necessary to maintain function (something that can be determined by "protein knock-out" experiments[23] and by establishing that no effective substitute is available in the flagellum's environment), then it follows that even some of the system's main parts exhibit irreducible complexity. It is hard enough to see how undirected processes could build a finely tuned machine by selecting and coordinating just the right major parts. But these processes would also need to build the main parts by selecting and coordinating just the right proteins. As we shall see, this is not a trivial task, because specialized proteins are typically incompatible with the majority of other available proteins and because the proper coordination of compatible proteins is vital to their successful implementation of a functional role.

The Indirectness Reply

Critics of Behe generally grant that the systems he calls "irreducibly complex" are too complex and too finely tuned to have been formed all at once. It would have required either a fortuitous "macromutation" that produced the system all by itself or the coincidence of many independent micromutations that happened to be perfectly coordinated to do the job. Allen Orr speaks for the majority of Behe's critics when he concedes, "Although this 'solution' yields a functioning system in one fell swoop, it's so hopelessly unlikely that no Darwinian takes it seriously."[24] Critics also concede that an irreducibly complex system cannot be built "directly." More precisely, such a system cannot develop from increasingly large, proper subsets of its unmodified parts by an incremental improvement in the same basic function. No such direct transition is possible because none of the proper subsets can perform the basic function. Instead, critics have returned to the central Darwinian idea of bait and switch. Instead of offering a direct explanation of the formation of these systems, they have offered a variety of indirect scenarios. These responses divide into two main approaches. According to one line, there are irreducibly complex biological systems (as defined by IC), but it turns out that they can be produced gradually after all. Thus it is claimed that irreducible complexity is not sufficient to refute Darwinism. According to the other main line, irreducibly complex biological systems do not exist. Contrary to appearances, all of these systems are reducibly complex.

Taking the first line, critics of Behe have focused on a loophole in his original definition of IC. They point out that the definition is compatible with an indirect gradualistic account of an irreducibly complex system.[25] When Behe says that removing one of the parts causes the system to "cease functioning," he is referring to the basic function mentioned earlier in the definition of IC. It is consistent with this that the remaining parts of the system can, either individually or collectively, perform some other function. Perhaps a mousetrap without a holding bar is an adequate tie clip, as Kenneth Miller has pointed out.[26] Certainly the spring could be used for many other purposes. Call the possibility invoked by the first line *decompositional polymorphism*, the ability of a single physical structure with one basic function to decompose into subsystems with different functions. A decompositionally polymorphic structure might need all five parts to function as a mousetrap but only four to play its role as a tie clip.[27]

Taking the second line, other critics of Behe dispute the existence of irreducibly complex systems in biology. They argue that the appearance of nonredundancy is an illusion. It has been argued that without some of its parts, the remaining parts of a mousetrap could be modified so that they alone perform the basic function of the whole system. The same function therefore develops indirectly via the different ways that the parts are adjusted to perform it. John McDonald has shown that one can have somewhat functional mousetraps with four, three, or two parts. And he has even proposed a one-part mousetrap![28] So perhaps the mousetrap exhibits redundancy after all. A related point is that even though a proper subset of the mousetrap's parts is unable to catch mice by itself, there may be resources ("spare parts") in a larger system that would support the formation of a perhaps less-efficient mousetrap. Although the system by itself is not redundant, it may be so in the context of its larger environment. Perhaps many interchangeable duplicate parts are readily available, or perhaps second-rate parts in the system's environment would allow the system to perform its function tolerably well until more specialized parts were developed. This would be like the child who builds a usable go-cart with odd parts that happen to be lying around in the garage, after which the child's father replaces some of the parts with others that are precision engineered. The various possibilities invoked by the second line can be summarized as *hidden redundancy*, the ability of a system to mask the availability of alternate, and in that sense, indirect, means of performing its basic function.

Both of these lines of response are developed in more detail in the following sections, where I also consider whether the idea of "scaffolding" provides a third alternative—that is, one system's collaborating in the construction of another. Here indirectness appears as one system's developing not from the ground up but via another support system. It will be shown that none of these variants of the "indirect" approach to overcoming irreducible complexity is a successful reply to Behe.

Decompositional Polymorphism: Irreducible Complexity Occurs in Biology but Does Not Refute Darwinian Gradualism

Co-optation

In Darwin's day the internal structure of the cell was completely unknown, and there was no science of biochemistry. However, Darwin did suggest that at the level of gross morphology co-optation might be a gradual means for building complex biological structures. Co-optation works by taking a structure with one function or no function and recruiting it to perform a new function:

> The illustration of the swimbladder in fishes . . . shows . . . that an organ originally constructed for one purpose, namely flotation, may be converted into one for a wholly different purpose, namely respiration.[29]

Likewise Stephen Jay Gould and Elisabeth S. Vrba distinguish adaptations from exaptations. Adaptations are selected for their current function, while exaptations are not. Exapted features "evolved for other usages (or for no function at all), and [were] later 'co-opted' for their current role."[30] Natural selection is thus presented as an opportunist who will exploit old materials if they happen to contribute to some current function. Kenneth Miller has recently defended this response to Behe. As a good illustration of how co-optation might build a complex biochemical system, he cites the metabolic process known as the Krebs cycle. This system makes use of parts that "could be selected separately, adapted first to different biochemical functions that have nothing to do with the eventual chemistry of the complete cycle. . . . [Yet] the individual parts . . . then are borrowed, loaned or stolen for other purposes."[31]

Exaptation is entirely plausible for biological functions that can be implemented by a variety of items, at least one of which is available for use. A sound analogy is given in one of Richard Scarry's children's books. A tourist, apparently from Texas, is exploring the great sea dyke on the coast of Holland. It so happens that there is a leak in the dyke. A nearby Dutchman is alarmed because he has no sandbag to plug the dyke. However, the tourist is available, and so the Dutchman recruits him to plug the dyke instead. Plugging a dyke is a simple task that all sorts of things can be appropriated to perform. One does not have to be specifically attuned to that task. Many items that currently have other uses or no use can be used to plug a dyke. Likewise, all sorts of things can be appropriated as doorstops: bricks, books, tennis rackets, college roommates, broken mousetraps, and the like.

Matters become more difficult when a function is highly specialized so that it is not likely that most items available for recruitment are competent for the task. To see how this can set limits on co-optation, consider an analogy with the roles of chess pieces. Let us suppose that each legal move of a chess piece represents a functional capacity, and let us stipulate that two pieces have the same functional capacities only if they are each capable of the same single moves, regardless of position. Thus my analogy assumes that only single moves represent selectable functions.[32] Now suppose we have the usual set of pieces but just one knight. If a given pawn is lost, then one of the remaining pawns can easily be co-opted to duplicate its role, since it has the same range of single moves. More interesting, the queen can cover for a lost bishop or castle since it is capable of all their single moves. But if the knight is lost, there is no one piece that can duplicate its role. The knight is the only piece capable of L-shaped moves. Consequently, the different functions of the other pieces are unsuitable to co-optation as surrogate knights. Thus, it is clear that co-optation has definite limits and is not guaranteed to provide an escape hatch for the Darwinist. If it turns out that the bacterial flagellum has parts analogous to the knight, which have no effective substitute, then co-optation is not an available option to defeat irreducible complexity.

I now wish to try to show in detail why co-optation is unlikely to account for the bacterial flagellum. Even at a high level of description, the flagellum has several highly specialized, well-matched, and interacting parts, namely, a paddle, a rotor, and a motor. For a working flagellum to be built by exaptation, the five following conditions would all have to be met:

- C1: *Availability*. Among the parts available for recruitment to form the flagellum, there would need to be ones capable of performing the highly specialized tasks of paddle, rotor, and motor, even though all of these items serve some other function or no function.
- C2: *Synchronization*. The availability of these parts would have to be synchronized so that at some point, either individually or in combination, they are all available at the same time.[33]
- C3: *Localization*. The selected parts must all be made available at the same "construction site," perhaps not simultaneously but certainly at the time they are needed.[34]

C4: *Coordination.* The parts must be coordinated in just the right way: even if all of the parts of a flagellum are available at the same time, it is clear that the majority of ways of assembling them will be nonfunctional or irrelevant.

C5: *Interface compatibility.* The parts must be mutually compatible, that is, "well-matched" and capable of properly "interacting": even if a paddle, rotor, and motor are put together in the right order, they also need to interface correctly.

Let us consider what it means for these conditions to obtain in the case of the bacterial flagellum. In a gradualist scenario, C1 can only obtain if the parts of the flagellum are formed independently, whether at the same time or a different time. Each of these parts would be nonfunctional, or they would perform a function different from the one they perform in the flagellum. Furthermore, none of the parts can be so tightly integrated with another structure, whether functional or not, that it is impossible for them to be co-opted for a new role.

The paddle is the simplest of the three elements. The filament "consists of a single type of protein, called 'flagellin.'"[35] It is therefore conceivable that something that has the right structure to serve as a paddle might be formed even though it has no function[36] or even if it has some different function. However, the paddle is of no use in forming the flagellum unless a rotor and motor are available, too. It seems wildly implausible that such highly specialized units as the rotor and motor would be available even though they serve either no function or a different function from the one they serve in the flagellum. Certainly it is a puzzle how these two parts could develop independently of each other and of the paddle, as gradualism requires. Just what does a rotor do without something to rotate and without a power source? If the rotor does nothing, why would such a specialized unit form? And if a nonfunctional rotor did form, why would so costly a waste of resources be retained? Likewise, it seems improbable that a motor specialized to work with the rotor would form, even if the motor serves a different function or no function at all. Indeed the idea that something so well suited to locomotion would form without its having a function is highly implausible. And even if it did form by some lucky mutation, it would be too costly in resources for natural selection to retain it. As Dawkins concedes,[37] natural selection cannot store up lucky mutations because they will be useful in the future; it can only preserve current function. So if the motor were available, it would have to serve some current function. But if, as we are assuming, that function is different from the function served in the flagellum, the motor's attunement to the former function would make it highly unlikely that the motor would also happen to work with the rotor (and paddle) and hence unlikely that it could be co-opted. That is, even if C1 (availability) were satisfied, it is highly unlikely that C5 (interface compatibility) would be satisfied as well. If the rotor and motor are compatible because they already form an integrated unit, we would need to know how such a system was formed (and retained) in the first place, given that it has no evident function with nothing to rotate. Furthermore, given the relative simplicity of the paddle, the problem of explaining how a suitable rotor and motor could form an integrated unit is almost as difficult as that of accounting for the whole flagellum.

It is also quite implausible that a rotor and motor serving a different function would be available for recruitment, since both would already be dedicated to service in some other system. If the rotor and motor are simply wrested from their current role, then this other system will likely malfunction. If the rotor and motor of a lawn mower are cannibalized to serve in a powered hang glider, then the lawn mower no longer works. Since we are assuming that the function from the ancestor system is different from the function of the successor system, it cannot be claimed that the new system will perform the old function as well—as if the hang glider swoops down to mow the lawn. Further, while it is possible that some less-than-critical functional structures might be dispensed with, there is also a danger that the older function had greater survival value so that a net reduction in fitness results.

This problem is only avoided if the ancestor system contains redundancy. This possibility is consistent with the resulting system's being nonredundant and even irreducibly complex. Pursuing our metaphor, we would need a lawnmower that has two rotors and two motors, even though only one of each is necessary, so that the extra rotor and motor could be donated to the hang glider.[38] Theoretically, this could happen in the biological case via gene duplication. However, complex redundancy is costly in resources and therefore unlikely to be retained by natural selection. Nor is it plausible that the same gene duplication would occur frequently enough to offset the atrophy of redundant features. Thus, it is quite unlikely that redundant copies would be available for co-optation. But even supposing they were available, it would be a remarkably improbable coincidence if redundant copies of parts attuned to one system worked correctly in a quite different system. It is rather unlikely that the rotor and motor of a lawnmower would serve in a powered hang glider without extensive modification and attunement to the hang glider's different requirements. And the same is to be expected with the flagellum, given its highly specialized parts. Natural selection cannot co-opt incompatible nonfunctional parts because they will be compatible and able to perform a function after modification. Thus even if C1 and C2 were satisfied, C5 seems insurmountable.

Nor will it do to suggest that a rotor and motor might somehow serve the functions of the ancestor system and the successor system. For one thing, there is no evidence of this, since the flagellum has a dedicated rotor and motor: the rotor and motor serve the single function of propulsion in the bacterium. But in any case, it is highly unlikely that the rotor and motor could be attuned to their highly specific current function at the very time that they are attuned to some different function as well. The motor and rotor of a lawn mower cannot be used to power a hang glider and mow the lawn at the same time.

Finally, explaining the formation of an irreducibly complex system by appeal to a redundant donor system only pushes back the problem of accounting for the origin of the former system's parts. If it is puzzling how natural causes could generate something like a powered hang glider, it is surely no less puzzling how they could produce a lawn mower, with or without an extra rotor and motor. The Darwinists surely have to concede this point; however, they might reply that the problem can be circumvented by supposing that a large number of donor systems, with each one's providing a single type of protein or at most a few such types. In that case, no one donor system need have the very same complexity as the system we are trying to explain. Might not extensive gene duplication provide a pool of extra proteins large and reliable enough to make it reasonably likely that all of the proteins needed to build the flagellum are available at the same time?

Let us grant that this is possible and thus assume that C1 and C2 are satisfied. The problem is that bacteria are capable of producing many types of proteins, and it is highly unlikely that undirected processes would be able to localize in a single construction site just those types required to build a flagellum. In other words, granted C1 and C2, it is still very unlikely that C3 is satisfied. Dembski demonstrates this point with the example of the bacterium *E. coli*.[39] In *E. coli*, the components of the flagellum are built by selecting several copies of 50 specific proteins from a possible 4,289 coded by *E. coli*'s genome. Given the vast number of types and possible combinations of proteins, what are the odds that just the right types are located together in the same construction site? Dembski provides a calculation of the localization probability based on conservative estimates of the resources:

> Let us . . . assume that 5 copies of each of the 50 proteins required to construct *E. coli*'s flagellum are required for a functioning flagellum. . . . We have . . . assumed that each of these proteins permits 10 interchangeable units. That corresponds to 500 proteins in *E. coli*'s "protein supermarket" that could legitimately go into a flagellum. . . . But those 500 reside within a "protein supermarket" of 4,289 proteins. Randomly picking 250 proteins and having them fall among those 500 therefore has probability $(500/4,289)^{250}$, which has order of magnitude 10^{-234} and falls considerably below the universal probability bound.[40]

What is more, this incredibly low probability only concerns the chances that undirected processes could get the right building blocks to the right place (C3). It does not take into account the need to coordinate the construction (C4) or the problem that many pairs of proteins have incompatible interfaces so that only a small subset of combinations is even possible (C5). Dembski calls the probability that all the parts can be correctly assembled without prohibitive, interfering "cross reactions" the "configuration probability."[41] Once this is factored in, there is no realistic chance that undirected processes could build the flagellum from the ground up.[42]

Even if undirected processes could account for the localization of the materials (C3), they are inherently incapable of the abstract top-down coordination required to manage the assembly of the proteins (C4). Natural selection is a bottom-up problem solver; that is, natural selection can only select or exploit components for their current function, not for some future function that they might have in a more complex system. Natural selection is blind and has no blueprint for the construction of a flagellum. As a result, the flagellum would have to be built by a series of lucky increments of functional complexity. The problem is that with a multipart system, a bottom-up process such as natural selection has no way of coordinating the development of each of the system's parts. Even if something that could serve as the flagellum's rotor did form gradually, it would be nothing short of a miraculous coincidence if, quite independently and with no anticipation of its connection with the rotor, a suitable motor developed that happened to be available at the very same time.

Further, even if these two components did develop independently, it remains mysterious how they would come to be finely attuned to each other (i.e., how C5 could possibly obtain as well). As Gould says, a certain amount of apparent "pre-adaptation" can be written off as "exaptation considered before the fact."[43] But this is surely implausible when two complex, specific parts must develop a finely tuned affinity for one another. First, how and why do they do this? Second, how and why do they do it when, until the point that the fine-tuning is successful, the system provides no function for selection to work on? The fact that each of the parts might have another function is irrelevant in this case. The other, independent functions of the rotor and motor cannot be improved by their increase in mutual compatibility since this only concerns their joint function. But the joint function can only be selected for after the increase in mutual compatibility reaches a certain threshold value, sufficient to support that function. So the increase would have to continue even though it initially provides no advantage, which is a thoroughly non-Darwinian preadaptation. What the rhetoric of exaptation conceals is the fact that parts with a given function are almost never going to be properly fine-tuned to play a different role in a highly specialized system. Furthermore, by its very nature, natural selection has no means of anticipating or selecting for that future role.

In some ways the idea of building a complex system like the flagellum by the contribution of multiple independent donors only intensifies the problem for the Darwinist. At the level of gross physiology, we know that without careful, intelligent choices of just the right donor organ for a given patient, rejection is highly probable. It turns out that exactly the same problem arises at the biochemical level. Proteins are not like basic Lego blocks, easily reusable and interchangeable. On the contrary, they have quite a low threshold of mutual compatibility: "For example, if it is the job of one protein to bind specifically to a second protein, then their two shapes must fit like a hand in a glove. If there is a positively charged amino acid on the first protein, then the second protein better have a negatively charged amino acid; otherwise the two will not stick together."[44] As a result of their high degree of specificity, most donor proteins are simply going to be rejected by a given "host" protein. It is not therefore a trivial matter to get just the right proteins to team up, any more than it is a trivial matter to find just the right donor organ for a patient. Most combinations fail, and in the case of donor organs, it is intelligent selection, not blind processes, that find the few compatible cases.

To summarize this section: Localization (C3), coordination (C4), and interface compatibility (C5) seem to pose insurmountable problems for cooptation, even if we grant the simultaneous availability of all the relevant parts of irreducibly complex systems (C1 and C2). I think one can see the severity of the obstacle for Darwinism quite clearly by an analogy with a theatrical production. It is easy to see that a typical stage play is an irreducibly complex system. A stage play needs multiple "parts," or actors, each of whom contributes a role. The parts must be available to perform at the same time. The parts are drawn from a large pool of possible contributors (actors) and are localized in one place, a theater. The parts are highly coordinated: there is a coherent narrative. They're interacting: the parts communicate. And they're well-matched: barring a foul-up, many of an actor's lines are the proper response to lines uttered by another actor. In reality, stage plays are produced by the creativity and top-down design of a playwright, and the actors must all conform to their proper role. Because coordination and attunement are built into the play itself, all the actors have to do is adjust their behavior to realize their largely preexisting roles. But might not unintelligent, bottom-up processes have generated the stage play?

In a plausible co-optation account, one that attempts to build the play from the ground up, we would have to presuppose that a lot of out-of-work actors are all available at the same time—not an implausible scenario. Somehow, from this pool, just those actors who are competent to perform the play are drawn together in the same theater. Perhaps the play is a Shakespearean tragedy, so the majority of actors, especially those appearing in mindless modern-day sitcoms, should be flatly rejected. The first problem for Darwinism is to explain how all and only the right actors end up in the same theater, ready to contribute to the same play. The problem is that the play does not exist yet, so any function it may perform cannot be the principle for selecting these actors. In a Darwinist scenario, actors do not know what the final play is about and thus fall into their roles by chance, quite independent from one another. As a result, most random selections of actors will probably produce bizarre "plays," such as the one with actors attuned to the roles of Count Dracula, Mrs. Doubtfire, Dirty Harry, Stuart Little, and Anne of Green Gables. It is fairly unlikely that any actor equipped to play a Shakespearean role would be included. But it is far less likely that only such actors would be found. The second problem is that, even given the right sort of actor, Darwinism has no evident means of coordinating the scenes. Darwinism is bottom-up, so each scene must develop incrementally and independently with no overarching plan or narrative. How do we get just the right actors to appear in just the right scenes? Even with actors exclusively geared to Shakespearean roles, if it is Macbeth and not Duncan who is killed early in *Macbeth*, the play falls apart. What is more, because each scene develops oblivious of the others with no top-down coordination, it is entirely probable that one scene would be from *The Tempest*, another from *King Lear*, yet another from *Coriolanus*, and still another from *Julius Caesar*. It is far more likely that Darwinian processes would build an uncoordinated medley of this kind than a coherent stage play. Finally, even if all the scenes belonged to the same play, Darwinism provides no evident means of adjusting the behavior of the actors so that it is well matched. Non sequiturs, pointless pauses, comments made to the wrong interlocutor (or a prop), and questions succeeding their answers would all be on the cards. The stage management would owe more to Monty Python's "gumbies"[45] than to the Royal Shakespeare Company. Darwinism would outperform existentialism in the theatre of the absurd.

Incremental Indispensability

Allen Orr is one Darwinist who concedes that co-optation is not a plausible explanation of irreducible complexity. As he points out, "You may as well hope that half your car's transmission will suddenly help out in the airbag department."[46] This perceptive comment shows that Orr realizes that an element tuned to one function is most unlikely to be tuned to a very specific, different function in another system. Preadaptation of such specificity would require foresight, precisely what natural selection lacks; however, Orr believes he has a more plausible alternative to co-optation, which Dembski has termed "incremental indispensability."[47] According to Orr:

> The logic is very simple. Some part A initially does some job (and not very well, perhaps). Another part (B) later gets added because it helps A. This new part isn't essential, it merely improves things. But later on A (or something else) may change in such a way that B now becomes indispensable. This process continues as further parts get folded into the system.[48]

Orr's description is vague, and it is hard to see how it could be applied to the flagellum.[49] More important, it does not seem to constitute a fundamentally different solution to others already proposed. Dembski rightly points out that the key question is whether the developing subsystem has the same function as the system itself or a different one: "If it is a function different from the one that the system ultimately attains, then we are really talking about a system formed by co-optation."[50] But Orr himself rejects co-optation. Yet, if the subsystem has the same function as the whole system, then by definition that system is not irreducibly complex. In that case, what Orr is really proposing is a form of hidden redundancy that argues in favor of simpler precursors with the same function as the whole system. That is the proposal I consider next.

Hidden Redundancy: Irreducible Complexity Does Not Occur in Biology

Internal Redundancy

John McDonald has attempted to cast doubt on the very idea of irreducible complexity. Taking the example of the standard five-part mousetrap, he has developed two series of functional precursors with fewer parts.[51] This approach suggests internal redundancy because it seems to show that the mousetrap's function can be performed by a proper subset of its own resources, with no help from outside donors. In every case, McDonald's strategy is to reconfigure the reduced set of parts so that they alone do what the larger set did, although perhaps not so well. Dembski has already given a thorough reply to McDonald,[52] so I will merely add a few further points that cast doubt on the relevance of McDonald's examples to Behe's thesis.

In McDonald's original series of transitions to a five-part mousetrap, each simpler system has its remaining parts modified and finely attuned to a different technique of trapping mice. Technically, one could argue that this does not conflict with irreducible complexity since, as Dembski points out, "irreducible complexity fails only if removal of one part *without* modification of other parts retains the system's original function."[53] But even if that requirement is waived, McDonald does not provide a plausible scenario for gradualistic evolution of the mousetrap and not merely because, as he readily concedes, his transitions are unlikely to mirror biological processes.[54] For any of McDonald's five mousetraps, if M_n is a successor mousetrap with n parts, with M_{n-1} its precursor with $n-1$ parts, the attunement of parts for M_{n-1} to its particular style of mouse catching actually unsuits it to be selected as a subsystem of M_n. M_n requires these same parts, plus an additional one, to be attuned in a quite different way, appropriate for a different style of mouse catching. For example, the way that the straight part of the spring is modified to serve as a rudimentary hammer in the two-part mousetrap (figure 4.3) unsuits it to engaging the hammer in the three-part mousetrap (figure 4.4). Likewise, the way the hammer is bent to compensate for the lack of a holding bar in the three-part mousetrap makes it unsuited to serve in the four-part mousetrap (figure 4.5), in which the hammer and holding bar are specifically adapted to one another.

In general, the fact that the parts of M_{n-1} are "well-matched" relative to its way of trapping mice actually prevents them from being well-matched for the way M_n does the same thing. It is not possible to generate a more complex, functional mousetrap simply by adding another part. Both the part added and the parts already present must be specifically adapted to one another to

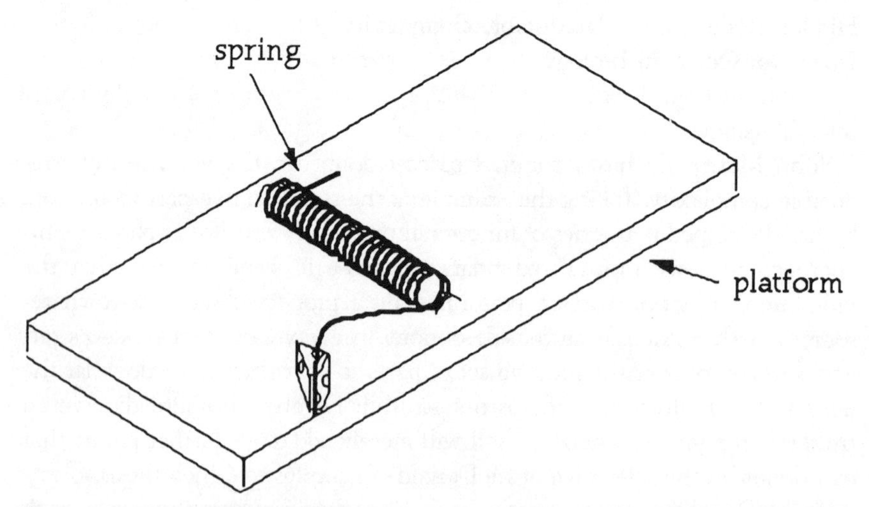

Figure 4.3. A two-part mousetrap. (Copyright © 2001 John H. McDonald. Reprinted with permission. All rights reserved.)

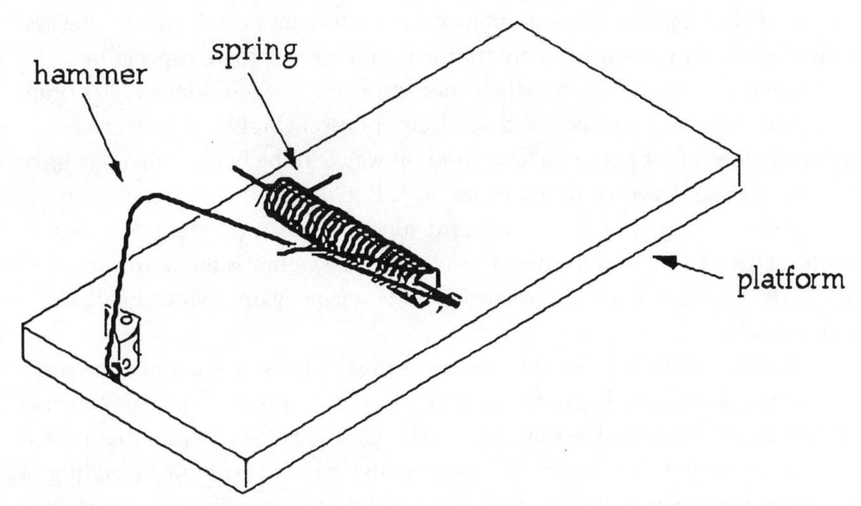

Figure 4.4. A three-part mousetrap. (Copyright © 2001 John H. McDonald. Reprinted with permission. All rights reserved.)

achieve function. In a similar biological case, this would appear to require not merely one mutation to generate a new part but a set of synchronized mutations that would both generate the new part and modify existing parts to suit. But Darwinists concede that this is far too improbable to be a reliable general explanation, even if it might occasionally occur. This objection is reminiscent

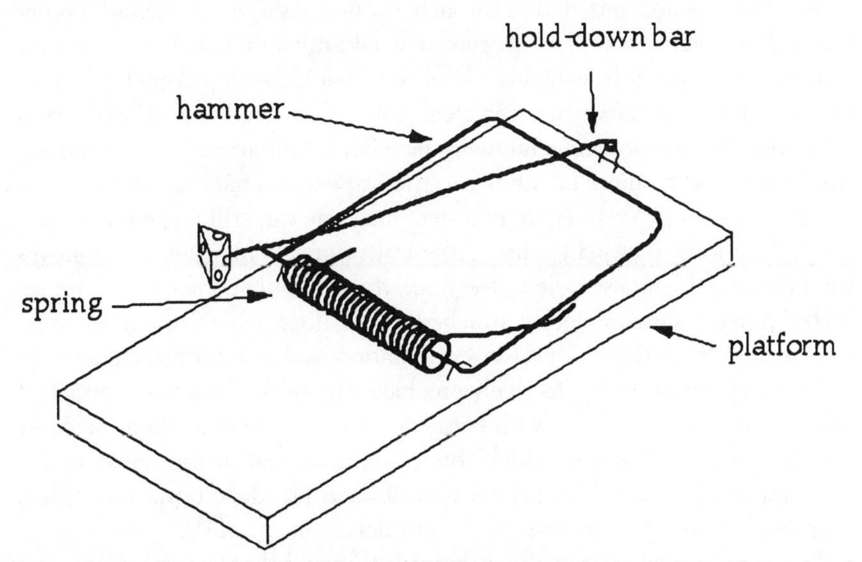

Figure 4.5. A four-part mousetrap. (Copyright © 2001 John H. McDonald. Reprinted with permission. All rights reserved.)

of one of the objections to co-optation: the attunement of a part to one system's function makes it unlikely that it would serve a highly specialized role in a different system. The parallel is not surprising because McDonald's original series reduces to a special case of co-optation. McDonald's scenario requires a system that performs function *F* in way *w* to be built from a precursor that performs *F* in some different way *w**. But this is just the special case of co-optation where it is not a different function but a different way of performing the same function that is co-opted. Thus, mutatis mutandis, many of the general objections to co-optation can be made against McDonald's original scenario.

But what about McDonald's second series? The series is more sophisticated and significantly more gradual, showing minor changes occurring within each *n*-piece mousetrap. McDonald goes so far as to claim that his examples show that "a complicated snap mousetrap can be built by adding or modifying one part at a time, with each addition or modification increasing the efficiency of the mousetrap."[55] McDonald's new transition starts with some spring wire positioned in front of a mouse hole, bent so that one end of the wire rests against the other. It gradually develops a coil and is then reoriented, becoming attached to the floor for stability. A movable platform appears, and this works even better. The incipient hammer grows longer and becomes L-shaped and then U-shaped, all the better to trap mice with. A holding bar appears, and because the original vertical part of the spring wire is no longer needed, it dwindles through atrophy. As a result, the last mousetrap is "irreducibly complex" in that it would not function if a single part were removed; yet, it is reducibly complex in that it developed gradually from functional precursors with fewer elements. To his credit, McDonald tries hard to avoid simultaneous mutually beneficial "mutations" in the mousetrap's parts, adding more credibility to the Darwinian character of the series.

Nonetheless, two replies are in order. First, one can still make some technical objections to at least some of the transitions. When the spring moves from framing the mouse hole to the floor, this will only confer an advantage if the spring is simultaneously attached to the floor, which requires two simultaneous "mutations": one for reorientation and one for attachment. In addition, this transition effectively precludes the next, since being attached to the floor is incompatible with being attached to a mobile platform. Most seriously of all, the way in which the spring is located in the center of the platform seems to smuggle in quite a bit of illicit preadaptation, since this is exactly the right place for the subsequent development of the trap.

However, it may be that this is hairsplitting and that such problems can all readily be fixed. For the sake of argument, then, let us suppose that Mc-

Donald is right and that the standard mousetrap can be developed gradualistically. My second point is that this does not show very much. It shows that the mousetrap is an internally redundant system in the sense that a proper subset of its parts can perform the same basic function. It does not show that all complex systems are internally redundant, because, by the nature of some mechanisms, several parts need to be independent of one another. All rotary machines need both a rotor and a stator. Since the stator must be stationary at the very time that the rotor is moving, these two functions cannot be performed by a single part. More generally, certain activities inherently require several distinct items. All effective swimming devices require a paddle, a rotor, and a motor. Thus, to swim at all, bacteria need something to play the role of each of these components. Furthermore, there is no evidence that these parts could all be simulated by a single item made of a uniform substance, such as the single piece of malleable wire in the one-part mousetraps. In the flagellum, the shape and function of the various items appear to require them to be built of distinct proteins, rather than by duplicating and reconfiguring the same kind of element. It is also an engineering necessity to distinguish the item being rotated and the source of that rotation. Even though the two parts might be integrated in a single unit, the engineering tasks are distinct and at some level would appear as discernible structural differences.[56] What is more, the environment of a bacterium is not swimmer friendly since an enormous amount of energy must be exerted to overcome Brownian motion, thereby allowing the bacterium to follow glucose up a nutrition gradient. There is no evidence that a bacterium could overcome Brownian motion with significantly less than the standard flagellar equipment found in existing bacteria—despite the fact that some flagella are slightly simpler than *E. coli*.[57] In this sense, there seems to be a "minimally complex" core of distinct elements that must be in place to perform the function of a flagellum, one that does not admit of further substantive simplification.

With generous concessions to his interpretation of Behe, we could allow that McDonald has shown that some systems that appear irreducibly complex, such as the standard mousetrap, are not. But he has not shown that this applies to all such systems, and in particular, he has not shown that the flagellum is reducibly complex.

External Redundancy

Unlike internal redundancy, external redundancy points out that simpler precursors of a system might be able to perform the same function with the help of outside donors. Perhaps if all government funding were withdrawn from National Public Radio (NPR), outside donors would enable it to continue its function. Thus, although NPR might seem to need government assistance to remain viable, that part of its fiscal structure is dispensable in the context of a large number of affluent, loyal listeners ready to step forward to sustain its influence.

We have already seen that the idea of an external system's donating spare parts to build a complex, finely tuned system is beset with considerable conceptual difficulties. However, such difficulties might evaporate in the face of clear empirical examples where what seems to be an irreducibly complex system is externally redundant. This is what Shanks and Joplin have argued, claiming that biochemical systems exhibit not irreducible complexity but redundant complexity.[58] They give the example of the metabolic pathway which results in glucose-6-phosphate. They point out that there are several different raw materials (enzyme isoforms) that can be used to synthesize this product and that, because of gene duplication, all of them are readily available. Indeed, one can prove their availability by knocking out the usual enzyme isoform: some other isoform takes its place and the same product is synthesized.

There could hardly be a clearer case of external redundancy than this. The environment of the pathway is so rich in compatible spare parts that it is false that removal of a part terminates the system's basic function. Hence, although it is complex, this particular metabolic pathway is not irreducibly complex. However, Behe has never claimed that all complex systems are irreducibly complex. And, as he points out in his reply to Shanks and Joplin, "The observation that some biochemical systems are redundant . . . does not show that they all are."[59] There are systems that are not redundant in Shanks and Joplin's sense. In some cases, it turns out that what is required is a unique biochemical, admitting of no substitute. For example, "Hexosaminidase A is required to catabolize ganglioside G_{M2}; its loss results in Tay-Sachs disease."[60] More generally, many dangerous and even lethal hereditary diseases are the result of a deficiency in a single naturally irreplaceable biochemical component, which is why so much research is being done on germ-line therapy to prevent these deficiencies. Clearly, those spared such hereditary diseases are blessed with irreducibly complex systems. Again, Behe has pointed out that mice without fibrinogen, tissue factor, or prothrombin are unable to form blood clots, making them liable to bleed to death.[61] Given such nonredundant systems, the challenge of irreducible complexity remains.

Cranes and Scaffolding: A Third Way?

We have seen that the two main lines of reply to Behe are quite unconvincing. But perhaps there is a third approach, one that exploits Dennett's intuition about the power of cranes. As mentioned in chapter 1, cranes are systems that enable the construction of buildings far more sophisticated than would be feasible in their absence. Likewise, scaffolding makes it possible to build an arch, even though in its intermediary stages the arch would collapse without the scaffold.[62] Let us designate cranes, scaffolding, and other systems that aid the construction of another structure, "support systems." The support systems themselves may be reducibly complex, built in gradual increments. But during the construction process, the support systems may be indispensable to the structure's development. While the support systems are in place, the structure being built may benefit from redundancy; that is, it remains supported even if parts are replaced. Yet, when the structure is complete, the support systems are redundant and may be removed from the site. When this occurs, the remaining system is irreducibly complex because removal of any part would cause it to collapse. Might there not be biochemical cranes or scaffolds that assist in building a system but that later disappear as costly redundancies?

How is this idea supposed to work in detail? Logically, there are three main options, all of which can be seen by considering the three possible answers to the following "support question" (SQ):

> SQ: What is the function of the support system before the appearance of the irreducibly complex system?

The least plausible answer is that the support system itself performs no function. But if so, then the support system would not have been selected in the first place, for the very reason cited in the removal of the support system after the structure is built. A support system cannot be selected for its usefulness in constructing some future function when it currently has no usefulness. This is to smuggle in teleology.

A more promising line claims that the support system performs some function *F* different from the function *F** performed by the irreducibly complex system. In this case the support system and the new system jointly exhibit decompositional polymorphism, and *F** has developed as a sort of parasite function that co-opts the support system as its host. In other words, this reduces to a subtle variant of the co-optation reply. However, this reply is vulnerable to several objections. First, if the support system's function *F* were important, then that system would surely remain, even after the construction, since the new system does not perform it. Thus, there is no reason to think that completion of the new structure would cause the support system to atrophy. But then this is not really a "crane" or "scaffolding" scenario since cranes and scaffolds do not stay around after the job is done. Second, and more important, there is no good reason to think that any parts attuned to serve function F^* in the new system would form. Any parts that appear would be selected for their current use, either serving F or some other function they can perform alone; they would not be selected for their future use, serving F^*. And the better attuned these parts were to serve F or another function, the less likely they would be attuned to serve F^*. So it would be highly unlikely that the support system would promote the development of a system that could perform F^*. If F^* is just a different function from F, it seems that F^* can only begin to be selected if all of the relevant parts are both available and coordinated at the same time. But this is far too improbable to be taken seriously. If this is right, then the system performing F is not really a "crane" or "scaffolding" for a system that performs F^*, since the whole point of cranes and scaffolding is to facilitate the gradual building of a complex structure. Thus, for both these reasons, if the support system performs some different function from the resulting structure, then the analogy with cranes and scaffolding breaks down.

The remaining possibility is that the support system performs the same function as the "irreducibly complex" system. Then, although that system may be irreducibly complex, its function is not, since before it disappeared the support system (a reducibly complex system) could perform it alone. This is an interesting combination of hidden redundancy and co-optation. The construction of a new structure co-opts another structure that performs the same function. It follows that until the support system disappears, the new, developing system is not irreducibly complex but externally redundant since the function would continue to be performed by the support system even if a part were removed from the new system. Only later, when the support system disappears, is the new system irreducibly complex since there is no longer anything in the new system's environment to ensure that the function is still performed if the system loses a part. In this form of hidden redundancy the redundant features are only present historically, rather than remaining in the environment as in Shanks's and Joplin's example.

Interestingly, this last possibility is one to which Darwin himself alludes in the *Origin of Species:*

> Two distinct organs sometimes perform simultaneously the same function in the same individual. . . . There are fish with gills or branchiae that breathe the air dissolved in the water, at the same time that they breathe free air in their swim-bladders. . . . In one of these cases, one of the two organs might with ease be modified and perfected so as to perform all the work by itself, being aided during the process of modification by the other organ; and then this other organ might be modified for some other and quite distinct purpose, or be quite obliterated.[63]

Let us ignore the fact that this example is from gross morphology, not biochemistry, and assume for the sake of argument that one organ of respiration is built with the help of another such organ in the way Darwin suggests. Let us also assume that the new organ is irreducibly complex.[64] The main problem with Darwin's example is that it never explains how either organ of respiration was formed in the first place. It only tells us how, given two such organs, one might be a support system for the improvement of the other. Matters are not helped if we suppose that initially there is only one organ that performs the function and that it aids in the construction of its successor. The question of how the predecessor organ acquired the function is left unanswered. If it required a support system as well, there is a clear danger of regress. Further, to use this strategy, the support system must be reducibly complex. This, however, cannot merely be asserted. It needs to be explained how the support system can be built incrementally. Thus, in the case of the flagellum, talk of support systems comes too cheap unless one can specify how something with the same basic function as the flagellum can be constructed gradually.

What is more, even if such a system arose, it is difficult to see how the emerging components of an ultimately irreducibly complex system would be selected. Let us suppose that a bacterium develops some primitive means of locomotion by gradual increments. Now suppose that a part of the new system appears, say the rotor. The problem is that the rotor has nothing to contribute to locomotion unless there is a paddle and a power source attuned to work with it. But it is just as improbable that the support system would happen to be equipped with these specialized parts as they would happen to be simultaneously available in the environment. The parts of the irreducibly complex system cannot be selected for their future function when they are all assembled. They do not have any evident function by themselves. So it seems they could only be selected if they aided the support system in performing its function. However, if the parts are attuned to aid the support system, it is improbable that they will also be attuned to work with each other to perform that function without the support system. This would be equivalent to moving from a five-part mousetrap to a three-part mousetrap. The three parts that are attuned to help the two-part support system perform its function will not be attuned to trap mice on their own. In consequence, the two-part support system does not really help the emergence of the three-part system. The very idea of a support system fails because the attunement of parts to work with the support system clashes with their future role. In fact, the support system is actually an obstacle, rather than an aid, to building the irreducibly complex system. Since a support system is by definition designed to help the construction of another system, this option turns out to be incoherent.

On examination then, none of the three logically possible ways of construing the idea of a support system explains irreducible complexity.

A Diagnosis of the Failure of Darwinism

Repeatedly, we have seen that even if gene duplication can make all the parts of an irreducibly complex system simultaneously available, Darwinism cannot provide credible solutions to the problems of coordinating these parts and ensuring their interface compatibility.

From my perspective as a teacher of computer programming, this limitation of Darwinism as a problem-solving strategy is unsurprising. First, consider the analogous problem of coordinating a program's instructions. As programs become more complex, it becomes virtually impossible to get them to work if they are written from the bottom-up, one instruction at a time. With so many details, it is highly likely that some critical task is specified incompletely or in the wrong order. To avoid such errors, programmers find it essential to use top-down design. Top-down design is a problem-solving strategy that begins with an abstract specification of the program task and then breaks it down into several main subproblems, each of which is refined further into its subproblems. This strategy is epitomized by such things as recipes, where the task is broken down into ingredients and utensils (initialization), the mixing and cooking of the ingredients (processing), and a specification of what to do when the dish is ready (finalization). The same approach is clear in the instructions to build "partially assembled" furniture, such as a bookcase. First, the assembly of the bookcase is reduced to its major tasks, constructing the frame, back, and shelves. Then each of these tasks is specified in detail. At every level, the order of the tasks is important; for example, the back and the shelves cannot be installed until the frame is complete. A quality top-down design is sensitive to the proper placement of tasks, ensuring that a given task is not omitted, redundantly repeated, or performed out of sequence. In this way, top-down design facilitates the proper coordination of problem-solving modules.

Unfortunately, natural selection cannot implement top-down design. Natural selection is a bottom-up, atomistic process. Tasks must be solved gradually, independent from one another. There is no awareness of the future function of the assembled system to coordinate these tasks. If even intelligent agents (experienced programmers) require top-down design to solve complex problems, it is tendentious to suppose that unintelligent selection can solve problems at least as complex without the aid of top-down design.

In fact, even with top-down design, programmers find that it is necessary to do two levels of testing to produce a functional program. One level, *unit testing*, tests the function of a module in isolation from the whole program. The other level, *integration testing*, ensures that when all the modules are assembled, they interact in such a way as to solve the overall problem. Both kinds of testing are needed: it is a fallacy of composition to argue that since all the parts of a system work, the assembled system will also work. Compare the following examples. Each football player is fit; therefore, the team will play effectively. Each brick is sound; therefore, the resulting wall will be strong. The conclusions do not follow because it matters how bricks and players are coordinated, and it matters whether they are compatible. Say that each player is fit but that the offense tries to score only when it has lost possession and the defenders try to defend only when they have possession: the team will be hopelessly uncoordinated. And if each player has a different play for the same circumstance, the team will suffer from incompatible elements. Likewise, if bricks are sound but are piled at random or are incompatible in size and shape, it will be impossible to build an effective wall.

Unfortunately, Darwinism commits precisely this fallacy of composition in the case of irreducibly complex systems. It has to suppose that the independent unit testing of atomic components (which natural selection provides) is a plausible way of coordinating and attuning those components for their combined role. But it is not. The majority of subsets drawn from the power sets of sound football players and bricks will be completely dysfunctional when combined as teams or walls.[65]

Matters are made worse for the Darwinist when we notice another parallel between computer science and biochemistry. In traditional non-object-oriented programming, the modules of a functioning program are not analogous to independent bricks, which are unable to communicate with one another. On the contrary, there is a very precise interface, or "signature," that specifies the parameters used in communication between modules. In most languages,[66] this interface is very specific so that there is exactly one correct invocation of a given module. The function call and the function thus exhibit an interface analogous to that between two binding proteins. When programs become complex, the parameter passing becomes so complex that it must be carefully designed. Darwinism, however, must build each module's implementation in blind independence of its possible communication with other modules in a multipart system. However, since two modules can only begin to communicate if all and only the right parameters are used, the correct communication procedure cannot be developed gradually. For the same reason, the sensitive and complex connection between certain proteins resists the kind of gradual development that Darwinism requires.

Conclusion

Behe's notion of irreducible complexity pinpoints an important obstacle to Darwinian accounts of biological function. His original definition seems vulnerable to a variety of objections. Some argue that although irreducible complexity exists, it can be produced gradually. Others argue that there is no such thing as irreducible complexity, because all plausible candidates exhibit hidden redundancy. Yet others suggest that the support system of a crane or a scaffold is the answer. As ingenious as they are, close examination shows that none of these proposals is convincing. Not only that, there are systematic theoretical reasons drawn from the study of other finely tuned complex systems, such as typical computer programs, that question the reach of Darwinism. Undirected bottom-up, atomistic processes are inherently incapable of the coordination and fine-tuning required by irreducibly complex systems. If this is right, then SAR and WAR cannot rely on Darwinism to explain the appearance of biological function, and all the reductionist projects that depend on this are thereby put in jeopardy. SAR cannot use Darwinism to provide an eliminative diachronic reduction of human agency, and WAR cannot use Darwinism to naturalize intentionality as a functional category of biology.

Notes

1. Richard Dawkins, *The Extended Phenotype*, rev. ed. (Oxford: Oxford University Press, 1999), 46.

2. Daniel Dennett, *Darwin's Dangerous Idea*, 220.

3. Michael Denton, *Evolution: A Theory in Crisis* (Chevy Chase, Md.: Adler and Adler, 1986), 229.

4. This is the line recently taken by Robert Koons in his *Realism Regained*.

5. A "mind-first" account is one that says that some mind is ontologically prior to matter so that it is not possible to account for all minds in materialistic terms. Dennett discusses mind-first accounts in the first chapter of his *Darwin's Dangerous Idea*.

6. Adaptionism tends to be given rather extreme definitions by its opponents, such as Lewontin, but the main idea is to assume as a working hypothesis that biological structures have some purpose that they perform optimally until the evidence shows otherwise. Antiadaptionists point to structures with no evident current purpose, arguing that they may be practically inevitable, given some other functional structure, or else vestiges of structures that had a functional role in the organism's ancestors.

7. Here "Darwinist" is intended in a broad sense, embracing the views of classical Darwinism and neo-Darwinism. In fact, many of the arguments in this chapter show the limitations of any undirected causation of the kind proposed by scientific materialism, whether it is strictly Darwinist or not.

8. However, a nonmaterialist like Koons, who allows for teleological causation, can help himself to whatever is valuable in Millikan's ideas. It is a mistake to think that only a materialist could hold that thought is a biological category. A theist might hold that human cognition reduces to biology but that biology implicates an intelligent designer who is not further reducible to anything else.

9. William Dembski offers a series of refinements to IC leading to a tighter definition that is more robust in the face of counterexamples yet also broader in scope. See his *No Free Lunch* (Lanham, Md.: Rowman & Littlefield, 2002), 279–85.

10. Michael Behe, *Darwin's Black Box: The Biochemical Challenge to Evolution* (New York: Touchstone, 1998; first published 1996).

11. To anticipate McDonald's objections (see the Internal Redundancy section, later in the chapter), I have to say that this is not to deny the possibility that different sections of a single piece of wire could implement both functions.

12. Dembski refers to such systems as "cumulatively complex" and gives the example of a city that can retain its function of community even with reduced population and services. See his *No Free Lunch*, 248.

13. This is a particularly plausible scenario because all that is required is duplication of the instructions for single stitching.

14. The converse does not hold, however. Any functional one-part system that lacks a backup is nonredundant, for example, a student's very last staple for the project completed at 2 AM, when no other staples are available; but the staple is not irreducibly complex, because it lacks multiple parts that interact and that are specifically adapted to one another.

15. Examples presented in detail by Behe in *Darwin's Black Box* include the cilium, the bacterial flagellum, the blood-clotting cascade, intracellular transportation systems, and the biochemistry of the immune system and metabolic self-regulation.

16. Even before Behe, the importance of this example as a challenge to Darwinism was noted by Michael Denton in his *Evolution: A Theory in Crisis*, 223–25. Behe discusses the example in his *Darwin's Black Box*, 69–72.

17. These statistics are from Dembski, *No Free Lunch*, 254.

18. Dembski, *No Free Lunch*, 288.

19. Indeed, any effective rotary device—such as a fan, a lawnmower, or a propeller engine for an aeroplane—must have a rotor, something rotated, and a stator; and these must likewise be "well-matched" and "interacting."

20. As we shall see, critics of Behe attempt to circumvent this point by allowing modifications of the parts or by allowing that precursor systems had different functions.

21. Rejected by Darwinism, "preadaptation" is the idea that nature is able to design useful features in anticipation of their future use. This would require some form of intelligent foresight that natural selection does not possess.

22. Dembski, *No Free Lunch*, 250.

23. Protein knockout experiments are used to determine if a biological structure can continue to perform its function minus a given protein.

24. H. Allen Orr, "Darwin v. Intelligent Design (Again)," *Boston Review*, December/January 1996–1997, 29. This article is available online at bostonreview.mit.edu/br21.6/orr.html (accessed March 13, 2002).

25. Behe himself concedes the possibility but denies that it will help the Darwinist. He writes, "Even if a system is irreducibly complex (and thus cannot have been produced directly) . . . one can not definitely rule out the possibility of an indirect, circuitous route. As the complexity of an interacting system increases, though, the likelihood of such an indirect route drops precipitously" (*Darwin's Black Box*, 40).

26. Indeed, I wore one of Ken Miller's "mouse clips" at the "Design and Its Critics" conference (organized by Bill Dembski and me), held at Concordia University, Wisconsin, in June 2000.

27. Swiss army knives are decompositionally polymorphic. The function of the knife as a whole is to serve as a basic toolkit. At the same time, each of its blades performs some more specialized function (basic knife, can opener, corkscrew, screwdriver, scissors, etc.).

28. In fact, as we shall see, McDonald now has two series of mousetraps. The original series is available at udel.edu/~mcdonald/oldmousetrap.html (accessed May 13, 2002). A newer, more sophisticated series is available at udel.edu/~mcdonald/mousetrap .html (accessed May 13, 2002).

29. Charles Darwin, "Difficulties on Theory," chapter 6 of *The Origin of Species* (New York, N.Y.: Random House, 1979), 220.

30. Stephen Jay Gould and Elisabeth S. Vrba, "Exaptation—A Missing Term," in *The Philosophy of Biology*, ed. David Hull and Michael Ruse (New York, N.Y.: Oxford University Press, 1998), 52–71, 55. The article was first published in *Paleobiology* 8, no. 1 (1982): 4–15.

31. Kenneth Miller, *Finding Darwin's God*, 151. Miller argues that this example shows how complex biochemical systems may be accounted for in Darwinian terms. However, he does not show that the Krebs cycle is irreducibly complex, so it is not a clear counterexample to Behe's main contention.

32. I thereby exclude cases where a single move of one piece can be simulated by several moves of another piece. This captures the fact that for something to be selected by co-optation to perform function *F*, it must currently be able to perform *F*, not merely a half or a third of what is required to perform *F*. Here it is important to distinguish being able to do *F* half as well and having only half of what is required to do *F* at all.

33. This does not assume that all the parts are developed at the same time. They could have developed asynchronously, but they must, at some point, all be available at once to form the new mechanism.

34. I owe this condition to Bill Dembski. It is an informal counterpart of his "localization probability." See Dembski's *No Free Lunch*, 291.

35. Behe, *Darwin's Black Box*, 70.

36. Perhaps a nonfunctional paddle is cheap enough in its use of resources, or else it is an unavoidable by-product of some other functional unit and so can be retained until a rotor and motor appear.

37. See the first quote at the beginning of the chapter.

38. Perhaps this would be like the case of double stitching, where a primary system performs a function tolerably well but the function is more reliably performed with a secondary system. Thus, we may imagine that a standard lawn mower appears somehow, cutting the grass tolerably well. Then duplication of some of its assembly instructions leads to an additional motor and rotor, making a more reliable lawn mower.

39. Dembski, *No Free Lunch*, 292–93.

40. Dembski, *No Free Lunch*, 293. In his earlier work *The Design Inference*, Dembski argues for a universal probability bound of 1 in 10^{150}, arguing that events that are less probable than this cannot reasonably be explained by chance—that is, events for which we have an independent specification, which in this case is simply the function we are trying to explain.

41. Dembski's "configuration probability" encompasses both "coordination" (C4) and "interface compatibility" (C5).

42. There is a quite technical estimate of these probabilities in Dembski's *No Free Lunch*, 294–302.

43. Stephen Jay Gould and Elisabeth S. Vrba, "Exaptation—A Missing Term," 64.

44. Behe, *Darwin's Black Box*, 53.

45. In the world of Monty Python (essential background for any philosopher), a "gumby" is an uncoordinated, inarticulate, unintelligent person who is virtually oblivious to his environment. Anyone who has seen the gumby version of Anton Chekhov's "The Cherry Orchard" has a good visual image of a Darwinist play.

46. H. Allen Orr, "Darwin v. Intelligent Design (Again)," 29.

47. Dembski, *No Free Lunch*, 257. Dembski replies to Orr effectively and in some detail, so I will keep my remarks brief.

48. H. Allen Orr, "Darwin v. Intelligent Design (Again)," 29.

49. Orr himself suggests that it may be impossible to reconstruct the biochemical pathways that lead to a given system. We should ask: What is the difference between a physical process that perfectly covers its tracks and one that does not occur? None, from a purely empirical standpoint. As Stephen Jay Gould once said about the lack of transitional forms in the physical record, evolution cannot always be going on somewhere else.

50. Dembski, *No Free Lunch*, 261.

51. The five original mousetraps can be seen at his website, udel.edu/~mcdonald/oldmousetrap.html (accessed May 13, 2002). They are also reproduced in Dembski's *No Free Lunch*, 262–65. McDonald's second, more sophisticated, series shows multiple modifications within each *n*-part mousetrap and is available at udel.edu/~mcdonald/mousetrap.html (accessed May 13, 2002). The new scenario combines reducible complexity with Allen Orr's suggestion of incremental indispensability. McDonald writes that "a part which may be optional at one stage of complexity may later become necessary due to modifications of some of the other parts."

52. Dembski, *No Free Lunch*, 261–67.

53. Dembski, *No Free Lunch*, 265.

54. McDonald says that his mousetraps "are intended to point out the logical flaw in the intelligent design argument; they're not intended as an analogy of how evolution works." See udel.edu/~mcdonald/oldmousetrap.html.

55. See the last paragraph of udel.edu/~mcdonald/mousetrap.html.

56. For example, one might engineer a fan whose blades gather solar power to rotate themselves. Nonetheless, there would be a structural difference between these blades and ones that do not gather their own solar power. To claim that this was a "simpler" fan because it did not require an independent power source would therefore be misleading.

57. As Dembski notes in reporting the professional literature on flagella, "These bacteria do not have substantially simpler genomes than *E. coli*. . . . The [flagellar] structures are very similar." See *No Free Lunch*, 294.

58. Niall Shanks and Karl H. Joplin, "Redundant Complexity: A Critical Analysis of Intelligent Design in Biochemistry," *Philosophy of Science* 66 (1999): 268–82.

59. Michael Behe, "Self-Organization and Irreducibly Complex Systems: A Reply to Shanks and Joplin," *Philosophy of Science* 67 (2000): 155–162, 160.

60. Behe, "Self-Organization," 160.

61. Behe, "Self-Organization," 161. Behe presents the case for the irreducible complexity of the blood-clotting cascade at more length in his *Darwin's Black Box*, chapter 4, and he defends it against the critique of Russell Doolittle in his essay "The Modern Intelligent Design Hypothesis," 174–77.

62. Dennett discusses cranes and scaffolds in some detail in his *Darwin's Dangerous Idea*, 212–20.

63. Charles Darwin, "Difficulties on Theory," 220.

64. In fact, as Dembski points out (*No Free Lunch*, 259), the assumption appears to be false, since respiratory systems such as lungs actually contain a lot of redundancy. Lung cancer patients may retain somewhat functional lungs despite the removal of large portions of a lung.

65. From another perspective, Darwinism is also guilty of the reverse fallacy, the fallacy of division. It argues that because a given "irreducibly complex" system has a function, it therefore must be composed of subsystems with the same or a different function. But by itself the flagellum's motor neither supports locomotion nor any other function.

66. Some languages do allow variation in the number and types of parameters used in a call to the same function. However, these languages are not a good model of biochemical processes. For example, in the blood-clotting cascade, it is crucial to viability that exactly the right number and type of "parameters" are sent at exactly the right time. Otherwise, the organism is liable to die, either of excessive bleeding because it cannot form an adequate clot or of a stroke because it cannot limit the clotting.

CHAPTER FIVE

~

The Alchemy of the Mind: Indirectness and Psychological Unity

> Each illusory self is a construct of the memetic world in which it successfully competes. Each selfplex gives rise to ordinary human consciousness based on the false idea that there is something inside who is in charge.[1]

> It is an excellent illustration of the mess that tends to result when models drawn from the physical sciences are drafted in without good reason to explain human behaviour.[2]

In the last chapter, we saw that Darwinism is incapable of accounting for the appearance of at least some biological functions. Specifically, the atomistic, bottom-up processes of natural selection cannot explain the coordination and interface compatibility exhibited by the parts of irreducibly complex systems. The two main schools of Darwinian psychology have an analogous obstacle.[3] As we shall see, both of these schools fail to explain the coordination and interface compatibility exhibited by at least some of an agent's thought and behavior.

One school of Darwinian psychology originated in the work of E. O. Wilson.[4] He proposed a new approach to the social sciences that attempted to account for human behavior in terms of natural selection. He dubbed the approach "sociobiology," although it is now often called "evolutionary psychology." In its most extreme form—genetic determinism—sociobiology claimed that virtually any statistical regularity in human behavior traced to our genetic heritage. Critics of genetic determinism pointed out that culture and tradition provide a means of influencing thought and behavior that is largely independent of genes. As Harvard geneticist Richard Lewontin says, it is not a good argument to "claim that since 99 percent of Finns are Lutherans, they must have a gene for it."[5] More generally, because of the way human behavior varies with environmental factors regardless of genes, it is clear that genes are not the only or even the decisive factor.[6] Some, such as Gould and Lewontin, conclude that the social sciences can never be fully subsumed under Darwinism. Others, such as Steven Pinker, defend a modified sociobiology by arguing that although culture is important for the content of many of our mental states, natural selection is crucial to understanding how our cognitive architecture works.[7]

Proponents of the other main school of Darwinian psychology also concede that genes have us on a much longer "leash" than genetic determinists had claimed.[8] They are nonetheless committed to "universal Darwinism," the idea that Darwinism applies to every aspect of life, including human cognition. Proponents of universal Darwinism suggest that genes should be supplemented with a cultural form of Darwinian replicator, most often called the *meme*.[9] While the proper definition of memes is much debated, the basic idea is that a meme is a discrete memorable unit, such as an idea, catchphrase, aphorism, marketing slogan, or jingle. Dennett suggests that we get a sense of the appropriate complexity of a unit meme from an assessment of the human capacity to remember and imitate: "the units are the smallest elements that replicate themselves with reliability and fecundity."[10] Meme complexes, or "memeplexes," are then built by association so that, for example, a variety of baseball memes all cluster together. Memeplexes are "groups of memes that are replicated together."[11] Memes are said to be Darwinian[12] because they exhibit variation, replication, and differential "fitness."[13] More specifically, the reshuffling and recombining of old memes create a constant variety of new memes; these memes are copied, sometimes imperfectly, via imitation; and some of these memes catch on better than others and thus become more widely distributed in the "memosphere";[14] that is, more copies of these memes are found in human brains and in auxiliary means of information storage, such as books and computers. Just as our genes are carried and transmitted by our bodies, memes depend on such "vehicles" as speech, signs, paper, electronic media, and brains themselves.

Darwinian psychology has many problems, some of which are addressed in chapter 6. In this chapter, however, I focus on the inadequacy of Darwinian reductionist accounts of the self. Proponents of both schools of Darwinian psychology claim to have shown that the unified self, as traditionally understood, is an illusion. However, we will see that there is a sound reason to posit such a self. The self helps to explain the cohesion of practical and theoretical reason and the synthetic unity of conscious thought and experience, neither of which Darwinian psychology can account for. In what follows I first present the case of Darwinian psychology against the traditional self. Then I offer a series of arguments against the Darwinian approach and in defense of the self.

The Abolition of the Self

According to the traditional, or common-sense, perspective of folk psychology (FP), the self exists as a unitary, reidentifiable subject of experience, reasoning, and action. Consider the following explanation of FP.

> FPE: Jack waived his hand because he saw his friend and wanted to greet him.

FPE purports to give reasons for Jack's action of waving his hand: Jack's seeing his friend and Jack's wanting to greet him. In any reasonable argument, whether it is theoretical or practical reasoning, the premises or reasons must be appropriately related to one another and to the conclusion. If the premises and conclusion only seem to refer to the same things but actually do not, then we have a fallacy of equivocation.[15] More specifically, when we give personal reasons for a personal action, it must be that the reasons and action are tied to one and the same person. To see this, consider the following substitution instance of FPE.

> FPE*: Jack waived his hand because John saw his friend and because Alice wanted to greet him.

Not only is FPE* not equivalent to FPE, it does not provide a general model for practical reasoning. What John saw and what Alice wanted could ultimately be reasons for Jack's action, but they are not Jack's personal reasons. Jack's personal reasoning is not "John saw my friend, and Alice wanted to greet him; so I will waive my hand," even if what John and Alice do causes Jack to have personal reasons to waive his hand. When it comes to Jack's personal reasons, the only thing that counts is what Jack sees and what Jack wants. More generally, when FP explains an action via a third-person account of an agent's first-person practical reasoning, it presupposes that the self who does the action is the same as the self who has the reason to act.

Both schools of Darwinian psychology challenge FP's picture of the self. These perspectives may allow that the traditional self is a practically indispensable fiction, but they deny that it captures psychological reality. Richard Dawkins expresses both schools of thought in his *The Selfish Gene*. Let us call the first school the *selfish gene theory* and the second school the *selfish meme theory*. Both schools adopt an indirect bait-and-switch approach to the self. The self is not taken as a given that requires explanation but as an appearance that has emerged gradually via indirect means. Not only is design in nature an illusory appearance, so is the traditional self.

The Selfish Gene Theory of the Self

In *The Selfish Gene*, Dawkins argues that we have the characteristics that we do because it serves the metaphorically selfish interest of our genes. In particular, our body and brain are "survival machines" or "gigantic, lumbering robots" programmed and built by and for the genes that inhabit them.[16] As it happens, the way natural selection works[17] is to preserve functional modules, and this applies to our physiological and cognitive features. What this means is that our bodies and minds are essentially bundles of "good tricks"—modules that have enhanced our fitness—but that there is no one "master" module that defines the physical or psychological identity of a human being. Steven Pinker has followed this strand in Dawkins's (and Dennett's) thought, and makes the following claim explicitly about the self:

> There's considerable evidence that the unified self is a fiction—that the mind is a congeries of parts acting asynchronously, and that it is only an illusion that there's a president in the Oval Office of the brain who oversees the activity of everything.[18]

According to both Pinker and Dennett, the mind is not a "Cartesian theater" in which a constant central observer monitors input and output; rather, it is a society of agents.[19] Indeed the mind is legion, for it is composed of many "mindlets."[20] To be sure, Pinker acknowledges the fact that our various agents must somehow be coordinated. They must have an executive controller because "no matter how many agents we have in our minds, we have exactly one body."[21] Without this assumption, it would be impossible to explain how the mind guides the body in exactly one direction. The untenability of the alternative is vividly portrayed by the movie *All of Me*, in which Lily Tomlin's soul takes over the right part of Steve Martin's body while he retains control of his left side: "He lurches in a zigzag as first his left half strides in one direction and then his right half, pinkie extended, minces in the other."[22]

However, all such considerations show is that the self is a useful "illusion that has come about because Darwinian selection found it expedient to create that illusion of unitariness rather than let us be a kind of society of mind."[23] While a circuit may allow one mental agent to trump the others at any given time, no one mental agent is always in charge. To return to our earlier example, suppose we call three of the agents in a given person's mind "Jack," "John," and "Alice." According to the Dawkins–Pinker theory, it is conceivable that something like FPE* could occur since the agents in charge of seeing, desiring, and acting might all be different from one another. The illusion of a unitary, enduring self results from the fact that only one of the three agents is in charge at any given time. Why does it always seem to be "me" who is in charge? Not because "me" denotes the same enduring agent but because "me" always denotes the currently active agent. "I" and "me" are even more indexical than we thought.[24]

The Selfish Meme Theory of the Self

Steven Pinker is a Darwinian psychologist firmly attached to the selfish gene school,[25] but he is quite skeptical of the claim that memes and cultural evolution in general are Darwinian.[26] Nonetheless, Dawkins, Dennett, and even E. O. Wilson all belong to the selfish meme school because they accept the need for units of cultural evolution that transcend the influence of genes. According to Dawkins's formulation:

> Examples of memes are tunes, ideas, catch phrases, clothes fashions, ways of making pots or of building arches. Just as genes propagate themselves in the gene pool by leaping from body to body . . . so memes propagate themselves in the meme pool by leaping from brain to brain via a process which, in the broad sense, can be called imitation.[27]

Just as our bodies serve the interests of selfish genes, so our minds are driven along by selfish memes: "When you plant a fertile meme in my mind you literally parasitize my brain, turning it into a vehicle for the meme's propagation in just the way that a virus may parasitize the genetic mechanism of a host cell."[28] Since memes are replicators in their own right, they need not serve the interest of genes. Thus one man's meme for martyrdom may inspire imitation and thereby succeed in propagating itself, even though martyrdom tends to reduce offspring and thus thwarts the interest of the man's genes.

Although meme theory grants cultural evolution a degree of autonomy from biological evolution, the upshot for the self is surprisingly similar. Instead of the mind's being a collection of modules, it is an assemblage of memes. On this view it is a mistake to think of the mind as some independent, recipient observer and processor of memes. For, as Dennett writes,

> a human mind is itself an artifact created when memes restructure a human brain in order to make it a better habitat for memes. . . . It cannot be "memes versus us," because earlier infestations of memes have already played a major role in determining *who or what we are*. The "independent" mind struggling to protect itself from alien and dangerous memes is a myth.[29]

This plurality of the mind is to be expected, given the distributed character of the brain's processing. Dennett argues that the brain has no single place where everything "comes together." Cognition does not occur in a single "theater" supervised by a Cartesian self; rather, it consists of "multiple drafts" distributed across the brain. Each draft is a partial narrative that processes some of the signals that affect the brain. Although these drafts may influence one another, all of the drafts are not synthesized into any sort of "metanarrative," because "there is no one place in the brain through which all these causal trains must pass in order to deposit their content."[30]

The distributed character of the brain and its parasitization by multiple memes lead Dennett to paint a Humean picture of the self. Hume famously denied that any such thing as a self was manifest in our experience:

> For my part, when I enter most intimately into what I call *myself*, I always stumble on some particular perception or other. . . . I never can catch *myself* at any time without a perception, and never can observe anything but the perception.[31]

However, like Dennett, Hume was not a complete skeptic about the self. He conceded that what we call the self is analogous to a republic:

> As the same individual republic may not only change its members, but also its laws and constitutions; in like manner the same person may vary his character and disposition, as well as his impressions and ideas, without losing his identity.[32]

Likewise, if by "the self" we mean a republican collection of entities—each of which is sometimes "in charge," but none of which is always in charge—then Dennett seems to allow that there is such a thing as a self. He says that a self is "an abstraction defined by the myriads of attributions . . . that have composed the biography of the living body whose Center of Narrative Gravity it is."[33] What he denies is a unitary, persisting self. According to Dennett, what occupies the "Center of Narrative Gravity" is not always the same draft.

Susan Blackmore is more skeptical than either Hume or Dennett. According to Blackmore, there is nothing, not even a republic, worth calling the self. What we call the self is no more than a specialized "memeplex," a "self-organizing, self-protecting structure that welcomes and protects other memes that are compatible with the group, and repels memes that are not."[34] One particular memeplex is the "selfplex," a particular collection of memes that is central to who we think we are, including our core convictions and preferences. Blackmore thinks that the correct conclusion to draw is not that the selfplex is a self but that the selfplex creates the illusion of a self that does not exist. All that really exist in the mind are various memes, some of which are central and dominant enough to create the illusion of a self as distinct from other memes. It is not that we are in control of certain memes but that some of them command and control our brain. In consequence, human agency, with its implication of personal design and authorship of ideas and actions, is an illusion. A striking consequence is that there is no interesting distinction between artificial and natural selection.

> We once thought that design required foresight and plan, but we now know that natural selection can build creatures that look as though they were built to plan when in fact there was none. If we take memetics seriously there is no room for anything to jump into the evolutionary process and stop it, direct it, or do anything to it. There is just the evolutionary process of genes and memes playing itself endlessly out—and no one watching.[35]

Against those who suggest that cultural evolution provides a person a means of resisting one's selfish genes,[36] Blackmore argues that there is no person but rather an ongoing, impersonal competition among memes and between memes and genes. By a different route, Blackmore is led to the same eliminative approach to human agency that I considered and rejected in chapter 2.

In Defense of the Self

Some Darwinian psychologists accept only the selfish gene theory of the self while others accept the selfish meme theory as well. Some Darwinists are also skeptical of the selfish gene theory in general (for example, "group selectionists"[37]), yet they might accept the selfish meme theory. However, the selfish gene and selfish meme theories do have some important commonalities. Both theories claim that our cognitive capacities are the indirect products of a multitude of autonomous, atomistic, blind, selfish replicators, governed by bottom-up Darwinian causal processes. I offer five considerations to show that it is precisely this feature of both theories that makes them psychologically inadequate.

Atomism and Alchemy

One of the strongest critics of sociobiology is Richard Lewontin, who argues that Darwinism tends to reify a model of social individualism, according to which groups are always determined by their component parts:

> With the change in social organization that was wrought by developing industrial capitalism, a whole new view of society has arisen, one in which the individual is primary and independent, a kind of autonomous social atom that can move from place to place. . . . This atomized society is matched by a new view of nature, the reductionist view. Now it is believed that the whole is to be understood only by taking it into pieces, that . . . the atoms, molecules, cells, and genes are the causes of the properties of the whole objects.[38]

The view is then transferred back to culture itself, inspiring E. O. Wilson to "search for the basic unit of culture."[39] The assumption is that culture itself must just be another substance composed of atoms, a "sack of bits and pieces such as aesthetic preferences, mating preferences, work and leisure preferences."[40]

The atomistic views of individuals, genes, and memes themselves trace back to the mechanistic philosophy and science of earlier centuries. According to this perspective, all causation results from the motion and impact of discrete particles. This position is cause for concern, however, because this mechanistic picture of physics has been thoroughly discredited. As noted in chapter 1, the propagation of electromagnetic waves does not require the mechanical medium of aether. Furthermore, as Werner Heisenberg pointed out, even the "elementary particles are not eternal: they can actually be transformed into each other."[41] What underlies this change is the continuity of mass energy, not anything identifiable as an atom. The idea that the atomistic view is simply the "scientific" view is false. It therefore cannot simply be assumed that an atomistic view of genes and memes must be correct or that the inability of such a view to account for the traditional self is reason to think that no such self exists.

What is more, there is independent reason to challenge the atomistic understanding of genes and memes. First, a combination of chemical atoms fully explains the resulting compound. But the combination of our genes does not fully account for our physical and psychological capacities. As Lewontin points out,

> it takes more than DNA to make a living organism. . . . A living organism at any moment in its life is the unique consequence of a developmental history that results from the interaction of and determination by internal and external forces. . . . Organisms do not find the world in which they develop. They make it. Reciprocally, the internal forces are not autonomous but act in response to the external.[42]

This is true even at the cellular level: "Part of the internal chemical machinery of a cell is only manufactured when external conditions demand it."[43] Lewontin gives the example of the enzyme that breaks down lactose, which "is only manufactured by bacterial cells when they detect the presence of lactose in their environment."[44] He has made much the same case against genetic theories of human psychological capacities, including his devastating critiques of the heritability of IQ.[45] Strong evidence suggests that external conditions, including nutrition and education, make a vast difference to psychological capacities, despite their independence from genes. Who we are and how we think is not simply a consequence of the combination of our genes. It also does not follow from evolutionary theory that our psychological capacities are adaptations that were selected for, as Dennett and Pinker seem to suppose. As Fodor has pointed out, if it is assumed that the human brain evolved gradually from an ape's brain, then the vastly superior psychological ability of the former would require a long and gradual series of changes resulting in a much more complex brain. However, at the anatomical level, human brains are very like ape brains; so from an evolutionary perspective, "relatively small alterations of brain structure must have produced very large behavioral discontinuities in the transition from the ancestral apes to us."[46] But if that is the case, cognitive capacities are not Darwinian adaptations that developed gradually. Either they are remarkable flukes of nature, by-products of changes that were selected for other reasons; or, as I believe, they require some nonnatural explanation.

Likewise, the combination of memes does not suffice to explain the coherent patterns of human thought. A coalition of atomistic, memorable units provides no basis for practical or theoretical reasoning. Humans can see certain thoughts and desires as reasons for further action or thought. However, memes are discrete units and are blind to their own and each other's existence.[47] Memes are not self-interpreting, nor are they able to interpret other memes. Consequently, a meme cannot see itself or another meme as a reason for some other action or thought. What is clearly required is an external interpreter of these memes. On pain of regress, this cannot simply be another meme or memeplex. The interpretive self cannot be reduced to a selfplex.

Second, atoms are supposed to be independent, contingently combining with other atoms to make molecules. But genes are related to other genes, both in sharing the same basic alphabet of nucleotide bases (A, C, G, and T) and, in some cases, by a similarity in the "meaning" of the instruction. It is not surprising that there are syntactic and semantic similarities between the instructions for assembling an insect's left antenna and those for assembling its right antenna. Likewise, memes share syntax and semantics with many other memes. The phrases "Just a moment" and "Just a minute" would both seem to be memes—discrete memorable units—but they are not independent because of their syntactic and semantic similarity. Likewise for the memes Darwinism, social Darwinism, and universal Darwinism. In fact, it is the norm for cultural ideas and movements to be seamlessly interconnected, which is, for example, why it is no surprise to see echoes of Hume's republic theory of the self in Dennett's account. The inherent interconnections between the content of these ideas is a much better explanation of this similarity between Dennett and Hume than the series of mental photocopying that memetic theory proposes. As Midgley says, in interpreting cultural patterns, we "need to look for the wider context of ideas out of which these patterns arise. . . . Explaining such things is primarily placing them on a wider map of other ideas and habits."[48]

What is more, atoms have their effects regardless of how they are interpreted. But the "syntax" and "semantics" of genes and memes make a difference to their effects. A genetic instruction has a certain "syntax" because the ordering of the bases matters: it makes a difference to the instruction's effects. Likewise, the instruction has a certain "semantics" because the cell "interprets" it as a blueprint for the construction of specific amino acids. Genes would not propagate in the way they do unless they had the right interpretation since this is required to produce useful features. Consequently, genes cannot be treated as uninterpreted atoms. Likewise, the effect of memes depends crucially on how they are interpreted by agents. "Just do it" may have been successful in promoting mindless hedonism, but fortunately it has not influenced those in control of nuclear missiles. Memes would not propagate in the way they do unless there were interpretive agents.

If neither genes nor memes really behave like atoms, we should be suspicious of an account of our psychological faculties that depends on the assumption that they do. The idea that what we call the self is thrown up by the instructions of selfish genes or is a conspiracy of selfish memes in a memeplex has very little plausibility if neither genes nor memes can be understood in isolation from their interpretation and wider environment. Further, atomistic accounts of integrated phenomena are often inadequate, as we saw in the last chapter. In particular, as I will now argue, the integrated character of an agent's reasoning cannot be explained by the bottom-up, atomistic processes of Darwinian psychology.

Reason, Practical, and Theoretical

A major problem for Darwinian psychology is that it cannot account for the ample evidence of psychological integration. One clear example of this is the human capacity for practical and theoretical reasoning.[49] An agent's reasons for action or thought are not discrete, unrelated elements. On the contrary, like the parts of the irreducibly complex biological systems we studied in chapter 4, these reasons are tightly coordinated and exhibit interface compatibility.

Suppose we take seriously the idea that the mind is the product of many independent selfish replicators (genes or memes) and that there is no unitary self. Then we should expect the kind of independent modular cognitive capacities that Dennett describes; in addition, if we accept memes, then we should expect these memes to be substantially independent of one another. However, both practical and theoretical reasoning issue in precisely one main conclusion—that is, the action of one body, or a particular belief or hypothesis held. To get just one conclusion, the reasons for it must be closely related to one another and to that conclusion.

First, consider practical reasoning. If Jack is going to produce the specific action of opening the refrigerator, this action can be explained if he desires a beer and believes that he can obtain one by opening the refrigerator. The belief and desire are coordinated so that the belief provides the means for satisfying the desire. The belief and desire together provide the means and motive for Jack's action. However, in a Darwinian mind, the beliefs and desires are independent atoms and may even develop in different narrative drafts so that there is no one place "where it all comes together." If so, there is no reason to think that a specific action will occur. If one part of Jack's brain contains a desire for beer (or suitable beer memes) and another contains a belief about the way to get beer (or suitable refrigerator memes) but the two do not come together, then the desire remains blind, having no means to satisfy itself; and the belief remains powerless because there is no motivation to pursue the means. For Jack to produce a specific action, it is required that the belief and desire are coordinated in one place. The belief is only a reason for action with the desire; likewise, the desire is only a reason for action with the belief. What is more, the belief, desire, and action must exhibit "interface compatibility." It does not work if Jack wants a beer and believes that he can obtain orange juice by opening the refrigerator. Nor does it work if Jack believes he can obtain a beer by opening the refrigerator but desires orange juice. In both cases, the reasons are not "well matched" and therefore do not jointly constitute a reason for Jack to open the refrigerator.

Furthermore, to return to the sort of example we gave at the beginning of the section The Abolition of the Self, it is clear that the reasons for Jack's action must, explicitly or implicitly, refer to Jack. If Jack desires a beer and Alice believes she can obtain one by opening the refrigerator door, that provides no reason for Jack to act. But who or what is Jack? Since the reasons must come together somewhere to result in action, it is natural to assume that "Jack" refers to an entity capable of surveying and interpreting all the reasons. This sounds like a perfectly valid reason to believe in a unitary self, inspecting the reasons on show in its Cartesian theater. The problem for the Darwinian psychologists is that they cannot give a plausible account of the reference of "Jack." If "Jack" refers to the module containing a given reason, then "Jack" will have one reference when speaking of his desire and another when speaking of his belief. But this will be just like "Jack's" desiring a beer and "Alice's" believing she can obtain one from the refrigerator. The single body that contains the two modules will have no reason to act. Even if "Jack" refers indifferently to both modules, there is no principle for uniting the reasons they contain in one act of reasoning. Something must see that the belief is a means to what is desired, which requires some single thing over and above the belief and desire. But the modules that contain the belief and desire do not constitute a unity, because there is nothing to tie them together. Likewise, if "Jack" is a selfplex, the problem is that the memes in the selfplex are unable to interpret themselves or each other as reasons. Distributed reasons cannot explain a specific action unless there is some interpretative agent who surveys them as a unity. The "cranes" of multiple drafts and memes obscure the facts about practical reasoning that the "skyhook" of the unitary self makes plain.

Many of the same observations apply to theoretical reasoning, although in this case the reasons need not make even an implicit reference to the bearer of those reasons. Suppose Alice argues that since $A = B$ and $B = C$, it follows that $A = C$. She will never reach this conclusion unless the two premises are coordinated in one place and their joint conclusion is drawn. Likewise, it is important that the reasons are "interface compatible." If Alice thinks that $A = B$ and $C = D$ or that $B = C$ and $A = D$, her reasons are not "well matched" and she will not (or should not) conclude that $A = C$. Furthermore, even given the right reasons, there must be some one entity ("Alice") that interprets these reasons together to account for the drawing of a specific conclusion.

It is perfectly clear, therefore, that human reasoning does require our thoughts to come together in one place and that it does require those thoughts to be well matched. There is no reason to think this will happen if our genes create independent processing modules, implementing separate drafts, or if memes are independent cultural atoms infesting our brain. Arguments do not simply form from the bottom up by the interaction of independent "drafts" or memes.

It is not at all coincidental that Darwinian psychology has the same difficulty explaining the unity and integration of human reasoning as Darwinian biology has explaining the unity and integration of irreducibly complex functions. Practical and theoretical reasoning is often irreducibly complex. A given argument has several well-matched, interacting reasons, and the removal of any one of them makes the argument break down. Attempts to build rational arguments by fortunate co-optation of available "drafts" or "memes" are most unlikely to work because an argument's reasons are so specialized to the tasks of interacting with the other reasons in the argument and of generating a specific action or thought. Appeal to redundancy does not help either. Although we do sometimes give redundant arguments, containing unnecessary reasons, we do not always do so. And if there is a supply of redundant reasons floating about, that still does not explain how just those that are compatible are coordinated to produce a coherent argument. The irreducible complexity of human reasoning points just as strongly to design as irreducible complexity in biology.

The inability of Darwinian psychology to account for human reasoning is devastating to its pretensions to be a science. The prestige of science depends on the application of highly advanced practical and theoretical reason. A "science" that is incompatible with such reasoning is therefore at odds with the very essence of scientific activity. In many ways, Darwinian psychology is an alchemy of the mind. As Dembski points out, the alchemists failed to provide a causally specific pathway from base metals to precious metals, but they did make vague gestures toward a variety of potions and furnaces.[50] What is more, excellent reasons emerged for thinking that the alchemists' proposed transformation was impossible. Likewise, Darwinian psychology does not provide a causally specific pathway from modules, drafts, and memes to human reasoning, but it does make vague gestures toward the magical powers of distributed processing and memeplexes. And again, there is excellent reason to think that no such transformation is possible.

Points of View

There is no coherent denial of the fact that human beings have points of view. To say one thinks that human beings have no point of view is itself to adopt a thought made possible only by having a particular point of view.[51] We have already seen, in chapters 2 and 3, that materialism is incapable of accounting for points of view because it tries to describe reality from an exclusively impersonal perspective, and the subjective view cannot be reduced to this perspective. If memes, or the modules constructed by genes, lack a real point of view, it is hard to see how one arises merely through their interaction. Atoms have no point of view, and when they combine, the resulting molecule also lacks a point of view. If modules or memes differ from atoms in this respect, we are owed an account of what this difference is. What is it about their combination that results in something as marvelous as a point of view? It seems most unlikely that such an account is forthcoming. Whatever else a point of view is, it is clearly something that enables the coordination of an agent's reasons for action or thought. But as we have just seen, this requires an observer or point of view that is independent of the reasons distributed across separate modules or memes and that holds them together in one place. The modules and memes themselves provide no such unity. Therefore, the memes and modules cannot account for such points of view.

There is also an obvious contradiction in Blackmore's claim that the self is an illusion and does not exist, even in a distributed sense. If there is no self and no such thing as a point of view, then there is nothing that can have the illusion that there is a self. To be deceived requires one to have a point of view. Deceive me all you wish, you will never deceive me into thinking I have no point of view.

However, suppose we bracket these objections and consider what happens if sufficiently complex modules or memeplexes do have points of view. Then there would be as many distinct points of view as there are such modules or memeplexes. Indeed, we should have to say, "My name is legion, for we are many." Clearly, however, psychologically "normal" people do not have the experience of such a divided mind, which is fortunate, as Pinker points out, since we have just one body to move around and we won't act effectively if this body receives independent, conflicting instructions. Pinker's solution is to posit a circuit that allows only one of the competing "agents" to be dominant. Perhaps we only have the illusion of a single enduring point of view because only one of the many competing points of view is active at any given time. Likewise Dennett might claim that only one draft is on the "front burner" at once and that is the one we identify with our point of view. In both cases it is assumed that the unitary self is an illusion. As I will now attempt to show, this claim cannot be sustained.

The Unitary Self

Let us suppose Pinker is correct and that there are many distinct points of view in our mind ($V_1, V_2, \ldots, V_n$) but that only one of the V_i ($1 \leq i \leq n$) is active at any one time.

We may suppose that these points of view occur in different modules or different memeplexes. In that case, who or what is having the illusion that there is a unitary self? The best reply seems to be that each of the V_i has this same illusion when it is active. Thus, suppose that at 4 PM, V_2 is active, and it has the illusion that it is the real self. Then at 5 PM, when V_{10} is active, it has the same illusion. Then it seems V_{10} won't notice that it was a different point of view, V_2, that claimed to be the real self earlier on. The obvious problem with this suggestion is that agents have the experience of remembering their past actions (and also plan for future ones). If V_2 plans an experiment and V_{10} conducts it, V_{10} will have to remember the plan V_2 formulated. More, to keep the illusion going, V_{10} will have to think of itself as having formulated the plan. But not only did V_{10} not formulate the plan, V_{10} is independent of V_2 and has no obvious way of knowing what V_2 did. How can V_{10} have the illusion of being the real self and of doing what V_2 did?

Perhaps, however, it is like relay, and V_2 passes on its memory of what it did to V_{10} like a baton. Then perhaps V_{10} will think that it planned the experiment. If V_2 thinks that it is the unitary self that planned the experiment and V_{10} inherits this memory, it seems V_{10} will now think that it is the unitary self that planned the experiment. Are we thus constantly engaging in revisionist history, claiming that we did things that were really orchestrated by a different point of view?

However, at least on standard accounts of personal identity, there is now a problem. Although I have reservations about the details of his account,[52] Derek Parfit has most fully expressed the widely held view that the identity of persons[53] depends on psychological continuity.[54] We know that our minds have different contents every day and that our access to our past thoughts and actions is incomplete. Nonetheless, if past mental states of a brain have appropriate causal relations to its present states, preserving important psychological characteristics, one may be willing to say that it was the same person then and now. In that case, the fact that V_{10} occurs in a different module (or in a different memeplex) than V_2 does not show that we have two different points of view or selves. In fact, on the assumption that V_2 does pass on relevant information to V_{10} like a baton, there is good reason to say that V_2 and V_{10} are the same points of view or selves since the baton preserves psychological continuity. But in that case, it is not an illusion but a psychological fact that there is a unitary self persisting through the thoughts and actions of V_2 and V_{10}.

Thus, the Darwinian psychologist who claims that the unitary self is an illusion is faced with the following dilemma. If each active module (or memeplex) has its own independent point of view, we cannot account for the experience of remembering past actions since the current module has no access to the information available to a past module. Yet, if the current module does have access to the information from the previously active module, then there is a mechanism for preserving psychological continuity; so, it is reasonable to claim that there is a persisting unitary self. Either the psychological Darwinists get the facts of memory wrong or they provide no compelling reason to deny a unitary self. Either way, their account of the self is mistaken.

However, it may be objected—rightly, in my view—that psychological continuity does not suffice to account for the unitary self. In the next section I consider what else is required, which will reveal another problem with Darwinian psychology.

Psychological Unity

The preservation of a unitary self over time is a diachronic problem concerning the relation between our past and present mental states. But there is also the synchronic problem of accounting for the fact that my present mental states one and all belong to me. Here, too, the Darwinian psychologist has a problem. The problem arises from the similarity between the views of Hume and Darwinian psychologists about the self.

Kant noticed a fundamental flaw in Hume's claim that the "self" is no more than a bundle or republic of experiences. The problem is that there is no principle of unity by means of which one can assert "I think" of all and only the experiences that belong to my (conscious) mind. Hume does not notice the significance of the fact that I can report many occasions on which "I enter most intimately what I call myself" and that this is what enables me to compare the experiences that "I always stumble on." If I can do this, then there must be some common feature of all these experiences in virtue of which I can remember and compare them. Without something over and above the many experiences to unite them as my experiences, "I should have as many-coloured and diverse a self as I have representations."[55] I would then be unable to compare my bouts of introspection because there would be no common self who experienced all of them. If Jack has one bout of introspection and Jill has another, then neither Jack nor Jill has any basis to compare both bouts. That is why there needs to be something that is identical in its relation to all of these experiences. Otherwise, each experience stands alone. If there are several modules or memeplexes active at once but no overarching principle, it will be impossible to explain how their experiences create even the appearance of being uniformly my experiences. Even if there are multiple points of view over time, we still need to explain how there can be only one point of view at a time, a point of view that unites various experiences in one consciousness. If it is claimed that there are many active points of view at the same time, we get a divided self that is not psychologically normal.

In this case, it won't help to claim that only one of these points of view is really active since the whole point is that all the experiences are going on simultaneously. There needs to be an explanation of the fact that subjectivity presents itself as a centrally focused point of view on the world, rather than a babel of unrelated points of view. Since at any given time, there is one such thing as my point of view, it cannot be identified with the point of view of any one of the simultaneously active modules. If one module contains my experience of coldness and another module contains my experience of wetness, there must be some point of view independent of either module that is able to judge that I am cold and wet. In fact the unity of the self is brought out by the very identity conditions of first-person thoughts. If Jack thinks to himself, "I am hungry," there needs to be something in virtue of which this is a thought about Jack and not somebody else. In other words, there needs to be something corresponding to the "I." As Quassim Cassam says in his critique of Parfit's reductionist approach to the self,

> The thinking of a first-person thought is essentially an event in the life of a subject, and *which* thought it is depends upon *which* subject's life it is a part of.[56]

This shows that psychological continuity is not enough to account for the unity of the self. If there is continuity between many independent drafts or memeplexes, there would be multiple independent selves, not one. It would be like a twelve-screen cinema with each screen showing a different movie and each of twelve observers able to watch only one at a time. This would provide twelve unrelated points of view, not one point of view that synthesized all these experiences in such a way as to be able to assert "I experience" of all of them. To be sure, there can be multiple, concurrent *un*conscious processes that are not synthesized in consciousness. But even in the case of conscious experience, there are numerous sources of experience that are somehow integrated into one perspective. I hear the birds, feel an itch in my scalp, and admire Cassam's philosophical insight all at the same time. To be sure, our conscious synthesis may be very incomplete; there may be selective attention, different degrees of awareness attending various items, and even editing. But that does not change the fact that the synthesis occurs and that something over and above the experiences synthesized holds them together as a unity. Yet again, there does need to be a place "where it all comes together" if we are going to explain the unity of conscious experience. And this seems like another good reason to posit a unitary self with its jolly old Cartesian theater.

There is also a practical reason to believe in unitary selves. As Christine Korsgaard has pointed out, from the perspective of practical reason, "You are one continuing person because you have one life to lead."[57] In particular, an agent who is embarking on a project (e.g., studying to enter law school) must think of oneself as being the one who (if all goes well) will be accepted by the law school; if accepted, that person will also think of oneself as having successfully implemented that plan. Otherwise, one is doing all that work to get someone else in school, and the person who gets in cannot feel proud of the accomplishment. Besides that, maybe the person who gets in did not want to go! It is also clear that our concepts of moral responsibility assume a continuing self, since we cannot praise or blame Alice for what Jane did. While such practical considerations alone do not prove that there is an enduring, unitary self, they do show that there is an enormous cost in denying it. Our plans and projects for our future lives become meaningless because it is someone else who collects, and quite possibly they never wanted to do it anyway. Our rewards and punishments go to the wrong person since it was someone else who acted well or badly.[58] Nobel laureates can only be awarded if time travel is developed and we can locate the proper recipient. Prisons are full of innocent bystanders who happened on the scene after a criminal in their body departed. We should require much stronger arguments than those provided by Darwinian psychology before leaping into such an abyss.

Besides, there is an alternative. According to intelligent design and the arguments of chapters 2 and 3, agency is an irreducible feature of reality. From this perspective, there is no problem in accounting for reasoning and for the unity of consciousness because there really is a place where it all "comes together." This enables intelligent design to uphold science by defending its use of practical and theoretical reasoning and to offer a better scientific account of the psychological facts about consciousness.[59]

Conclusion

The two schools of Darwinian psychology are motivated by an atomistic paradigm that has been rejected by physics and that does not work in biology or psychology either. The irreducible complexity of practical and theoretical reasoning cannot be explained by Darwinian psychology; instead, it points to design. Since the activity of science itself depends on such reasoning, Darwinian psychology is antiscientific. Further, our thoughts, experiences, and actions are synthesized and interpreted in a way best explained by positing an enduring unitary self of precisely the kind that Darwinian psychology denies. By making room for agency as an irreducible feature of reality, intelligent design is therefore in the right position to uphold the rationality of science and to account for the psychological facts.

Notes

1. Susan Blackmore, *The Meme Machine* (Oxford: Oxford University Press, 1999), 236.

2. Mary Midgley, *Science and Poetry* (London: RKP, 2001), 76.

3. Note that one can be a Darwinist and still be highly skeptical of Darwinian psychology. This position is defended by Jerry Fodor in chapter 5, "Darwin among the Modules," of his *The Mind Doesn't Work That Way* (Cambridge, MA: MIT Press, 2000).

4. See E. O. Wilson, *Sociobiology: The New Synthesis* (Cambridge, MA: Harvard University Press, 1975).

5. Richard Lewontin, *Biology as Ideology: The Doctrine of DNA* (New York: HarperCollins, 1991), 94.

6. For an excellent statement of the case against old-style sociobiology, see Tom Bethell's "Against Sociobiology," *First Things*, January 2001.

7. See Steven Pinker, *How the Mind Works* (New York: Norton, 1997).

8. Wilson himself has migrated to this school. See the chapter "From Genes to Culture," in Wilson's *Consilience: The Unity of Knowledge* (New York: Knopf, 2000). Wilson urges us to "search for the basic unit of culture" (134) and speaks favorably of Dawkins's suggestion that "memes" are such units, although he opts for a narrower definition (136).

9. The term *meme* was coined by Richard Dawkins in his *The Selfish Gene*, rev. ed. (Oxford: Oxford University Press, 1989). See chapter 11, "Memes: The New Replicators." Less popular terms for the unit of cultural evolution include "mnemotype, idea, idene, . . . sociogene, concept, culturgen, and culture type" (E. O. Wilson, *Consilience*, 136).

10. Dennett, *Darwin's Dangerous Idea*, 344.

11. Blackmore, *The Meme Machine*, 19.

12. Some object that, strictly speaking, memes are not Darwinian but Lamarckian, since memes pass on acquired characteristics. I take it, however, that the project of universal Darwinism only requires replicators that are broadly Darwinian in character and that memes are only intended to be partially analogous to genes. This is what Susan Blackmore argues (*The Meme Machine*, 59–62). In any event, for the purposes of this discussion, I will grant that memes are "Darwinian" in a nontrivial sense.

13. See Dennett, *Darwin's Dangerous Idea*, 343.

14. The term is coined by Daniel Dennett by analogy with "biosphere." His point is that there is competition to occupy the finite number and capacity of brains just as there is competition for scarce biological resources. See Dennett's *Consciousness Explained*, 206.

15. For example, consider the following argument: (1) all men are rational; (2) no woman is a man; so, (3) no woman is rational. This has the form of a valid argument but is invalid because in (1), "men" means "human beings," whereas in (2), "man" means "adult male."

16. Richard Dawkins, *The Selfish Gene*, rev. ed. (Oxford: Oxford University Press, 1989), 19–20.

17. Or the way it is supposed to work, since I deny this very point in chapter 4.

18. Steven Pinker, "Is Science Killing the Soul?" 12.

19. See Daniel Dennett, *Consciousness Explained* (Boston, MA: Little, Brown, 1991), ch. 5.

20. The term *mindlets* is used by Richard Dawkins, "Is Science Killing the Soul?" 14.

21. Pinker, *How the Mind Works*, 144.

22. Pinker, *How the Mind Works*, 145.

23. Dawkins, "Is Science Killing the Soul?" 14.

24. "I" is said to be an indexical because it does not have a constant reference; rather, it refers to its speaker or writer. If the Dawkins–Pinker theory is correct, "I" is still more indexical because it refers to the currently active agent; so, even in the mouth of the same body, and even over a fairly short period without any interference with the brain, "I" can refer to many different things.

25. Pinker writes that "the gene's-eye view predominates in evolutionary biology and has been a stunning success," and he defends Dawkins's *The Selfish Gene*. See Pinker, *How the Mind Works*, 43–44.

26. See Pinker's critical discussion of meme theory in his *How the Mind Works*, 208–10.

27. Dawkins, *The Selfish Gene*, 192.

28. Dawkins, *The Selfish Gene*, 192.

29. Dennett, *Consciousness Explained*, 207.

30. Dennett, *Consciousness Explained*, 135.

31. David Hume, "Of Personal Identity," *A Treatise of Human Nature*, ed. Ernest C. Mossner (New York: Penguin Books, 1984; first published 1739 and 1740), 300.

32. Hume, "Of Personal Identity," 309.

33. Dennett, *Consciousness Explained*, 426–27.

34. Blackmore, *The Meme Machine*, 231.

35. Blackmore, *The Meme Machine*, 242.

36. Remarkably, Dawkins himself claims that we can rebel against both genes and memes. He concludes *The Selfish Gene* with these words: "We are built as gene machines and cultured as meme machines, but we have the power to turn against our creators. We, alone on earth, can rebel against the tyranny of the selfish replicators." However, Dennett would surely be right to object that a fellow materialist is here appealing to a skyhook.

37. Group selectionists deny that the gene is the sole unit of selection. They argue that a group character may be selected, even if it hurts the chances of the genes of a particular organism in a group. Thus, a crow who raises the alarm to warn of an approaching predator risks its own genes by calling attention to itself, but it benefits other crows and, perhaps as a side effect, birds of other species in the same ecological community.

38. Lewontin, *Biology as Ideology*, 11–12.
39. Wilson, *Consilience*, 134.
40. Lewontin, *Biology as Ideology*, 14.
41. Quoted in Mary Midgley, *Science as Poetry* (London: RKP, 2001), 63.
42. Lewontin, *Biology as Ideology*, 63.
43. Lewontin, *Biology as Ideology*, 63–64.
44. Lewontin, *Biology as Ideology*, 64.
45. For a full statement of the case against the heritability of IQ, see Richard Lewontin, Steven Rose, and Leon Kamin, *Not in Our Genes: Biology, Ideology, and Human Nature* (New York: Pantheon Books, 1984). For a brief statement, see "All in the Genes?" chapter 2 of Lewontin's *Biology as Ideology*. Lewontin argues that "there is not an iota of evidence of any kind that the differences in status, wealth and power between races on North America have anything to do with genes, except, of course, for the socially mediated effects of the genes for skin color" (*Biology as Ideology*, 36).
46. Fodor, *In Critical Condition*, 209.
47. Later on, I bracket this objection and see what follows if I suppose that some memes or memeplexes do have points of view.
48. Midgley, *Science and Poetry*, 73.
49. In the arguments that follow I am going to simply assume that folk psychology and intentionality are substantially well founded, since I already argue for this at length in chapters 2 and 3.
50. Dembski, *No Free Lunch*, 239–43.
51. This is not a pun on "point of view." I do not mean that thinking there are no points of view is a point of view in the sense of an opinion. I mean that having that or any other particular thought requires there to be a subject who holds that thought. Rocks, seas, and anthills do not have thoughts because they are not subjects of thoughts and in that sense have no point of view.
52. I do not believe that Parfit's reductionist program, of showing that the self can be fully accounted for in impersonal terms, is successful because, as Quassim Cassam has argued, to be a self requires one to have a concept of one's own temporal extension. This concept makes ineliminable reference to a particular person. See Quassim Cassam, "Reductionism and First-Person Thinking," chapter 13 in *Reduction, Explanation, and Realism*, ed. David Charles and Kathleen Lennon (New York: Oxford University Press, 1992), 361–80.
53. For Parfit, a "self" may be only a part or a stage of a "person." However, I will use the terms "self" and "person" interchangeably.
54. See Derek Parfit, *Reasons and Persons* (Oxford: Clarendon Press, 1984).
55. Immanuel Kant, *The Critique of Pure Reason*, trans. Norman Kemp Smith, second impression (London: Macmillan, 1933), 154.
56. Cassam, "Reductionism and First-Person Thinking," 375–76.
57. Christine Korsgaard, "Personal Identity and the Unity of Agency: A Kantian Response to Parfit," *Philosophy and Public Affairs* 18, no. 2 (spring 1989): 101–32, 113. David Shoemaker attempts to show that Korsgaard's view is actually compatible with

Parfit's. See his "Theoretical Persons and Practical Agents," *Philosophy and Public Affairs* 25, no. 4 (autumn 1996): 318–32. Regardless of whether Shoemaker is right, he concedes the important point that practical considerations do shape the concept of an agent over time.

58. For a strong critique of the ethical implications of Darwinian psychology, see Phillip Johnson's "Darwinism of the Mind," chapter 5 in his *The Wedge of Truth: Splitting the Foundations of Naturalism* (Downers Grove, IL: IVP, 2000).

59. This point is reinforced in the next chapter.

CHAPTER SIX

Beyond Skinnerian Creatures: A Defense of the Lewis–Plantinga Argument against Evolutionary Naturalism

> Inference itself is on trial: that is, the Naturalist has given an account of what we thought to be our inferences that suggests that they are not real insights at all. We, and he, want to be reassured. And the reassurance turns out to be one more inference (if useful, then true)—as if this inference were not, once we accept his evolutionary picture, under the same suspicion as all the rest.[1]

> If all that exists is Nature, the great mindless interlocking event, if our own deepest convictions are merely the by-products of an irrational process, then clearly there is not the slightest ground for supposing that our sense of fitness and our consequent faith in uniformity tell us anything about a reality external to ourselves. . . . If the deepest thing in reality . . . is a Rational Spirit and we derive our rational spirituality from It—then indeed our conviction can be trusted.[2]

In the last chapter, we argued that two prominent Darwinian accounts of the mind are deeply flawed. Both accounts fail to explain the coordination and interface compatibility exhibited by an agent's practical and theoretical reasoning, and both are unable to capture the synthetic unity of conscious experience and judgment. In this chapter, I pursue a closely related concern, that Darwinian psychology fails to explain the reliability of our cognitive mechanisms. If the Darwinian account of the formation of these mechanisms is correct, there is good reason to think that our intentional states would be false or inappropriate more often than not and that our inferences would be even more prone to error than they are actually are. This undercuts evolutionary naturalism, the view that such processes as random variation and selection[3] account for all aspects of living beings, including their cognitive mechanisms.[4] In fact, as both C. S. Lewis[5] and Alvin Plantinga[6] have argued, it shows that evolutionary naturalism is self-refuting. According to Plantinga's formulation, if evolutionary naturalism were true, the probability that our cognitive mechanisms are reliable would be either low or inscrutable, taking away our justification[7] for supposing that any of our beliefs is true, including the belief in evolutionary naturalism.

Unsurprisingly, not everyone would accept these conclusions. Some, such as Dennett and Quine, assume that the mind is an adaptation, and they make the strong claim that evolutionary naturalism explains the reliability of our cognitive mechanisms. Others, such as Fodor, who doubt that the mind is an adaptation, make the weaker claim that evolutionary naturalism is at least compatible with such reliability. I begin by reviewing these attempts to defend evolutionary naturalism. Then I argue in favor of two important theses advanced by Alvin Plantinga. The first thesis warns of epiphenomenalism, which is the view that, at least in virtue of their content, propositional attitudes such as beliefs have no causal role.[8] According to Plantinga:

> A1: "It is exceedingly hard to see . . . how epiphenomenalism . . . can be avoided, given [Evolutionary Naturalism]."[9]

Then I support thesis A1 by developing two Lewisian arguments—one empirical, the other conceptual—connecting evolutionary naturalism and epiphenomenalism. If epiphenomenalism is true, then it is far more likely than not that our cognitive mechanisms are unreliable. Suppose, however, that a more substantive role for intentional states did evolve, along the lines suggested by folk psychology. Plantinga's second thesis is

> A2: Even granted that Folk Psychology is correct about the causal role of intentional states, "the fact that my behavior (or that of my ancestors) has been adaptive . . . is at best a third-rate reason for thinking my beliefs mostly true and my cognitive faculties reliable."[10]

Next, granted evolutionary naturalism, I support thesis A2 by arguing that psychological instrumentalism, the view that our beliefs can be useful without being true, is far more probable than psychological realism, the view that our beliefs are (normally) useful because they are true.[11]

What is more, evolutionary naturalism also implies theoretical instrumentalism, the view that scientific theories are merely useful computational devices: the output of these devices mirrors observable phenomena, but the theoretical models may be thoroughly fictional. Granted theoretical instrumentalism, one has no reason to think that science is a reliable means of discovering the truth about the fundamental categories in the universe. Evolutionary naturalism leads to the unwelcome conclusion that science has no authority to tell us what kinds of things belong in our ontology. This means that the rational motivation for pursuing science—the discovery of true accounts of what exists in the world—can no longer be justified. Once again, a commitment of scientific materialism undermines the rationality of science. In the final section, I outline what an alternative to scientific materialism might look like.

Evolution and Reliability

I will not attempt to give a rigorous definition of "cognitive reliability," but it helps to clarify widespread intuitions about its meaning. Consider an analogy between a cognitive mechanism and a radio station. Suppose that the radio station is government controlled and that half or more of the information it conveys is false propaganda. Then certainly the radio station would be an unreliable source of information. Reliability comes in degrees, yet it seems wrong to say that a radio station that tells the truth 30 percent of the time is 30 percent reliable. If a radio station communicates truth exactly 50 percent of the time but one has no independent means of determining which broadcast statements are true, it is simply a matter of chance whether one can acquire true beliefs by listening to the radio station. In that case, agnosticism, in the sense of withholding belief that a given broadcast message is true or false, is the rational response. However, if significantly less than 50 percent of the broadcast is true and one has no means of sifting out the true statements, then skepticism, in the sense that one believes a given broadcast is more likely false than true, is justified. In that sense, if a radio station tells the truth 30 percent of the time, then, assuming one cannot tell which statements fall into this 30 percent, one has good reason not to rely on the radio station at all. So while a radio station that tells the truth 80 percent of the time is more reliable than one that is accurate 60 percent of the time, reliability as we understand it has a well-defined metric only when more than 50 percent of the information is correct.[12]

One limitation of this analogy is that it only concerns the accuracy of information and hence only applies directly to the belief-forming mechanism.

This mechanism is reliable only if it has a significant chance of producing accurate representations—true beliefs. However, desires can be reliable or unreliable, yet they are neither true nor false. Rather, a desire-forming mechanism is reliable only if it yields appropriate desires, desires for the good. While "the good" certainly cannot be reduced to what promotes the survival of an individual creature,[13] we can nonetheless suppose that "appropriate" desires are frequently attuned to avoid danger, seek nutrition, and so forth. Since practical reasoning involves both beliefs and desires, it seems best to define cognitive reliability in a way that is sensitive to both kinds of representation. This suggests the following intuitive, necessary condition for cognitive reliability.

> CR: A cognitive mechanism M is reliable for an agent S only if, under normal conditions for M's functioning, M's deliverances to S are more often true (or appropriate) than not.[14]

This necessary condition is all we will need in the discussion that follows.

The evolutionary naturalist who prizes cognitive reliability in the intuitive sense suggested may defend either a strong or a weak thesis. According to the strong thesis:

> ST: Natural selection explains why human cognitive mechanisms are reliable, since true beliefs and appropriate desires are more fitness enhancing than false beliefs or inappropriate desires.

According to the weak thesis:

> WT: Natural selection is compatible with the reliability of human cognitive mechanisms, in the sense that natural selection does not make it probable that these mechanisms are unreliable.

In the following sections, I briefly consider how ST and WT have been defended, and I outline the case for thinking that the supporting arguments are inadequate.

The Strong Thesis (ST)

According to ST, there is a selective advantage in having reliable cognitive mechanisms. The assumption is that cognitive mechanisms are adaptations that enhanced creatures' ability to survive and reproduce. The argument is that false beliefs tend to lead creatures astray and hence make it less likely that they will survive. If a creature often sees its fellows being attacked and devoured by lions but forms the belief that lions are safe and should be welcomed, that this creature is unlikely to survive for very long. In Quine's words, "Creatures inveterately wrong in their inductions have a pathetic but praiseworthy tendency to die before reproducing their kind."[15] Although he only holds WT, Fodor thinks that proponents of ST have a measure of common sense on their side:

> What we all believe is that when actions out of false beliefs are successful, that's generally a *lucky accident*; and correspondingly, that a policy of acting on false beliefs, even when it works in the short run, generally gets you into trouble sooner or later.[16]

Likewise Dennett argues that "evolution has designed human beings to be rational, to believe what they ought to believe and want what they ought to want."[17] Dennett's formulation notes that it is not only beliefs but also desires that track our environment. It is easy to suppose that a creature that believes that lions are unsafe but who wants to be devoured will not be around for long.

However, ST makes at least two tendentious assumptions. First, ST assumes that evolutionary naturalism is compatible with our beliefs and desires having a significant causal role in directing our behavior. The truth of beliefs and the appropriateness of desires can be adaptive only if intentional contents are visible to natural selection. If all that matters is adaptive behavior and what we believe and desire cannot be selected for, then there is no reason to expect our beliefs to be true or our desires appropriate. If a creature believes that lions are safe or wants to be devoured but runs away anyway, it will survive. This point will be developed more carefully in the next section, Epiphenomenalism and Skinnerian Creatures. Second, ST assumes psychological realism, the view that our beliefs and desires are useful guides to behavior because they tend to reflect reality. However, for any given topic, it can be shown that there are vastly more systems of false beliefs than systems of true beliefs that produce the same behavior. In consequence, even if the content of intentional states is causally relevant to behavior, it is far more likely that natural selection would favor psychological instrumentalists, creatures with useful but frequently false beliefs and useful but frequently inappropriate desires, not psychological realists. This point is developed more fully in the section entitled Folk Psychology and Instrumentalism.

The Weak Thesis

Fodor supports Darwinian biology but has qualms about ST because he thinks that it is very likely that many of our psychological capacities are not adaptations but merely by-products of changes in the brain that were selected for other reasons:

> Since psychological structure (presumably) supervenes on neurological structure, genotypic variation affects the architecture of the mind only via its effect on the organization of the brain. And, since nothing at all is known about *how* the architecture of our cognition supervenes on our brains' structure, it's entirely possible that quite small neurological reorganizations could have effected wild psychological discontinuities between our minds and the ancestral ape's.[18]

As a result, Fodor does not expect natural selection to explain cognitive reliability but simply to be compatible with it. In fact, Fodor argues that even if our cognitive mechanisms are adaptations, it is still wrong to claim that natural selection explains their reliability:

> Darwin would have been just as happy if what had been an offer for selection to choose from when we evolved was cognitive mechanisms which produce, by and large, adaptive *false* beliefs. . . . There is . . . *no* Darwinian reason for thinking that we're true believers. Or that we aren't.[19]

For Fodor, all we can say is that reliable cognitive mechanisms somehow emerged. Whether these mechanisms are adaptations, natural selection does not explain their availability.

However, Fodor faces the same basic problems as the proponent of ST. First, his account still assumes that scientific materialism is compatible with the emergence of intentional states that are not epiphenomenal. Second, Fodor's nonadaptationist position makes psychological realism a fluke. Surely the systematic way that our beliefs and desires track appropriate behavior and external reality requires a better explanation than saltation. It is very implausible to claim that psychological realism is true merely as a result of changes in our brain made for other reasons.[20] On the contrary, the way our minds mirror action and the world is an example of tightly specified complexity. The problem is particularly acute because psychological instrumentalism is a priori far more likely than psychological realism. Indeed, design is a better explanation of psychological realism than either the chance proposed by WT or the interplay of chance and necessity proposed by ST, that is random variation and selection.

Epiphenomenalism and Skinnerian Creatures

A trait can be selected only if it is visible to selection. Selection preserves characters that better suit an organism to its environment. In the case of action, it is only the behavioral response that is directly visible because it is only the response that can be tested for its aptness to the environment. If any mental state results in that behavior, it is not directly tested. Yet, it is a fallacy to conclude that mental states cannot be selected. Mental states might still be selected because they do in fact produce adaptive behaviors. However, if adaptive behavior is all that ultimately matters, the interesting question is whether it is likely that the intentional content of these mental states is causally relevant to the behavior. I argue that the answer is no. If we presuppose evolutionary naturalism, then natural selection has no plausible means of getting beyond epiphenomenalism.[21] But, if epiphenomenalism is true, then the content of our intentional states makes no difference to our behavior; so it is highly unlikely that our beliefs are true or that our desires are appropriate.

Why suppose that thesis A1 is true, that evolutionary naturalism tends to epiphenomenalism? Two reasons: First, as we saw in chapter 3, attempts to naturalize intentionality fail—either via a synchronic functionalist account or via a diachronic derivation from natural selection. If naturalism does not have the resources to generate intentionality, it does not really explain the existence of intentional states. If in fact such states do exist, as anyone who wants to defend their reliability must grant, it follows on naturalistic assumptions that they must be an accidental by-product of some natural process. Since the only plausible candidate for such a process is natural selection, we must suppose that intentional states arose due to changes in the brain that were selected for other reasons, which is Fodor's position. However, if intentionality is not an adaptation, the naturalist has no clear reason for supposing that our intentional states are either accurate or appropriate more often than not.[22] If behavior is an adaptation but thought is not, the most likely result is that humans can produce adaptive behavior independently of thought—that is, epiphenomenalism. That natural selection is logically consistent with cognitive reliability does not affect the fact that natural selection makes epiphenomenalism much more probable.

Second, natural selection is a mechanistic process. An examination of the limits of this process shows that it is unable to account for the rationality of intentional states. But if thought is not rational, then it need not track behavior and the world in expected ways; so again, there is no reason to think that intentional states are reliable. In making this argument, it is helpful to use Dennett's classification of four kinds of creatures, each more sophisticated than the last.[23] According to Dennett, it all began with basic Darwinian creatures, replicators that are refined by random mutation and selection. Those with only hardwired responses do not do as well as those with enough plasticity to learn via random trial and error and operant conditioning—Skinnerian creatures. The limitation of Skinnerian creatures is that some trials are fatal errors, preventing further learning. So Dennett suggests that the next advance is Popperian creatures, creatures that are capable of "preselection among all the possible behaviors or actions, weeding out the truly stupid options before risking them in the real world."[24] Finally, Gregorian creatures emerge, equipped with such "mind tools" as words.

Let us grant for the sake of argument that natural selection can account for the emergence of Skinnerian creatures. This seems plausible because behavioral responses are visible to natural selection, and Skinnerian creatures are favored precisely because their responses are better adapted to the environment than those of non-Skinnerian creatures. However, I do not think that the same kind of explanation accounts for the emergence of Popperian creatures. Popperian creatures have "a sort of inner environment" with "lots of *information* about the outer environment and its regularities,"[25] and this environment is used to determine which actions are plausible to try. Here Dennett makes two enormous assumptions: first, that Popperian creatures have somehow acquired intentionality, since they have a mental model that represents the world; second, that Popperian creatures are able to reason in the abstract about the possible consequences of their actions. In chapter 3, I noted the problems with the first assumption. Here I want to focus on the difficulties raised by the second assumption. There are two main arguments for thinking that the emergence of rationality is not explained by natural selection. These arguments were suggested a long time ago, by C. S. Lewis in his book *Miracles*.[26]

The Empirical Argument for Epiphenomenalism

The first argument is empirical, resting on the observation that a creature's responses can be improved indefinitely without its acquiring rational thought. Because algorithmic improvements can be made in a creature's response mechanisms, the creature is able to "anticipate" novel conditions without its having to either think about its behavior or reason out its consequences. As Lewis says, "Such a perfection of the non-rational responses . . . might be conceived as a different method of achieving survival—an alternative to reason."[27] That this is possible has been shown repeatedly by work in artificial intelligence (AI). Sophisticated game-playing programs are equipped with heuristics that "direct" their behavior. If a given move leads to a loss, the program does more than a Skinnerian creature. A Skinnerian creature would simply reduce the probability of producing that move if the same circumstance occurs in the future. By contrast, using backtracking techniques, a sophisticated AI program can "see" which heuristics led to that bad play and, using metaheuristics, adjust the heuristics themselves.[28] Likewise, if a given move leads to a win, it is not only that move but also the heuristics that generated it that are more likely to be employed in the future. Because these same heuristics are used to generate many different moves, the AI program has the ability to generalize what it learns. When a novel situation arises, the heuristics to handle it will already have been modified by their use in similar situations. In that sense, the heuristics "anticipate" the new situation. That this really works is shown by the startling improvements in the general playing performance made by AI games programs. It is not the case that these programs simply avoid the same mistake twice or repeat good tricks: there is a global improvement in their game-playing strategy. Yet it is not plausible to claim that these programs are reasoning. The fact is that although the programs implement rational principles—the initial heuristics and metaheuristics used to improve them—their rationality plays no causal role in the output of the device.

Now consider a Skinnerian creature who happens to have some intentional states. Call this creature "proto-Popperian." We cannot assume that the proto-Popperian creature's intentional states simply happen to obey norms of rationality: that would be a vastly improbable coincidence, and how these states come to be rationally configured is precisely what needs to be explained. We can assume, however, that the Skinnerian mechanism does a fairly good job of producing adaptive responses and that it does happen to implement a somewhat rational principle for learning—"repeat what has rewarded you; refrain from what has hurt you"—despite the fact that it does not think. Selection can only operate on what currently works. Since the Skinnerian mechanism works, natural selection can select and refine it. Since the intentional states available to the proto-Popperian are not yet rationally configured and since improvements in the implementation of rational principles can be made without rational thought, it seems both redundant and highly unlikely that intentional states would be recruited to improve the Skinnerian mechanism.

The problem is particularly acute because Darwinian selection requires gradual improvements of functional elements, but intentional contents appear unable to contribute to the improvement of the Skinnerian mechanism that is already functioning. In fact, we have another example of the problems of coordination and interface compatibility (see chapters 4 and 5). How can randomly organized intentional contents be co-opted by a tightly specified Skinnerian mechanism all the while retaining functionality? Further, it would seem that the only way intentional contents might contribute something that could not more easily be done automatically is if those contents were under the control and supervision of an agent. But then how can intentional contents be combined with a nonintentional mechanism that is independent of that agent? The requirements for intentional contents to contribute anything new seem to guarantee that those contents will not be compatible with a Skinnerian mechanism.

In general, it is hard to see how one can simply add thoughts to an unthinking mechanism. It seems far more likely that these intentional states would either be redundant (since their contribution is already automated or can easily be automated) or deleterious since, being irrationally configured, they would make the resulting system less rational and functionally coherent than it was before. Thus, it seems highly unlikely that the intentional states would ever come to be configured rationally. But if they are not configured rationally they would have to be epiphenomenal; otherwise, they would conflict with the rational principles implemented by the enhanced Skinnerian mechanism, causing a substantial net reduction in fitness. In that case, natural selection would work to exclude the causal influence of intentional content on behavior since such influence would be a positive liability. Thus, there is sound reason to think that the thought of proto-Popperian creatures would remain epiphenomenal indefinitely. A really rational thinker, such as Popper himself, would never emerge.

The Conceptual Argument for Epiphenomenalism

The second argument for A1 goes deeper: natural selection is conceptually incapable of generating rational thought as a result of major contrasts between the process of natural selection and the process of reasoning.

Behavior is selected because it leads to advantageous consequences. Running away from lions is selected because it increases a creature's chances of staying alive. A Skinnerian creature is able to learn this fact by associating lions with danger. As a result, running away becomes a conditioned response in the presence of lions. Now suppose that a proto-Popperian creature—call him Karl—goes a step further and actually thinks about the matter. Perhaps what Karl thinks is something like *Lions mean danger* and *When I have run away from danger, I have stayed alive*. There is now an expectation that future lions will be dangerous and that running away is the correct response. Such an account might succeed in explaining the origin of primitive inductive inferences. But not only would it fail to explain how we see that such an inference is correct, it also cannot explain our grasp of deductively valid reasoning.

First, an expectation that the future will resemble the past is not the same as a rational inductive inference. As Lewis says:

> The assumption that things which have been conjoined in the past will always be conjoined in the future is the guiding principle not of rational but of animal behavior. Reason comes in precisely when you make the inference "Since always conjoined, therefore probably connected" and go on to attempt the discovery of the connection.[29]

In other words, even if Karl starts to think about his Skinnerian associations, they remain expectations and not fully logical inferences, unless he has some means of justifying the expectation. As Hume showed, there is no possibility of justifying this expectation based solely on the facts of the past and our experience of them. The ability to see that an inductive inference is correct depends on grasping its justification: the natural selectional history and Skinnerian conditioning of a creature, even supplemented with the proto-Popperian ability to represent that conditioning, cannot explain the ability to see that such an inference is correct. Induction is a reasonable principle only if human beings are reliably attuned to real, natural categories and only if there is an objective basis for their preference for theories that are "simple," "symmetrical," "beautiful," and so forth. The historical interactions between an environment and its inhabitants (whether a species or an individual) can never provide such a justification.[30]

Strict Popperians will not be bothered by the last point since they deny that inductive inference can be justified and so deny any such possibility as "seeing" that an inductive inference is correct. However, they do agree with most everyone else that deductive reasoning is justified and that this is something we know. This presents an even bigger hurdle for Karl. The encounters of Karl's ancestors and of Karl himself with the environment have nothing to do with the ability to see whether deductive reasoning is valid. To take Lewis's example, consider the simple inference $A = B$, $B = C$, therefore $A = C$. The fact that those of Karl's ancestors who thought this way had a better chance of survival does not explain why the inference is valid or how Karl could see that it is. The most that this can do is explain why Karl has a genetic tendency to think in this way. But all that this means is that Karl has a certain expectation that the inference will hold. It does not explain the ability to see that the inference necessarily holds. When one sees that an argument is valid, one sees that, given the premises, the conclusion must follow; but this can never be justified by the past success of an argument, which only shows that it has held. Thus the fact, if it is a fact, that had the inference to $A = C$ not been valid, it would not have helped Karl's ancestors, is irrelevant. The fact that reasoning that happened to be logical enhanced survival does not show why the conclusion necessarily follows. Again, if Karl himself happens to think through the inference to $A = C$ not because of his genetic heritage but because of a learned association between A, B, and C, that does not explain why it is a universally correct inference or how Karl could know this. All Karl knows is that the inference has worked for him and perhaps for others, which is compatible with its breaking down tomorrow. Yet this is not true of us. As Lewis says, "My belief that things which are equal to the same thing are equal to one another is not at all based on the fact that I have never caught them doing otherwise."[31]

I think we can see some common limitations of natural selection and Skinnerian reinforcement that explain their failure to account for our reasoning abilities. Both natural selection and Skinnerian reinforcement are undirected, contingent processes that operate solely at the factual level of cause and effect and a creature's actual responses. However, reasoning is directed toward its conclusion, has modal force, and obeys norms: it concerns what a creature ought to think or do. When Karl concludes that A = C, all we can say is that this is what he is genetically equipped or developmentally conditioned to expect. There is no ground for saying that Karl sees that A = C follows. This is because "follows" in the logical sense is quite different from "follows" in the psychological sense. When a thought follows psychologically, it simply means that, as a matter of contingent fact, that thought comes into one's mind as a causal consequence of other thoughts or ideas with which it is associated. The correctness or otherwise of this association has nothing to do with it. By contrast, when something follows logically, this means that there is an objectively necessary connection between certain propositions, whether or not anyone thinks them. That some propositions are directed toward certain conclusions in this way is a fact entirely independent of what people think or expect. Justified logical thinking is the grasping of an objective dependency of propositions, not the mere subjective enjoyment of thoughts that happen to exhibit that dependency. What is more, the dependency grasped is not contingent or psychological but necessary and independent of minds. And one sees the conclusion not as something one does think or has a tendency to think but as what one should think. Logic is normative, governing what a rational agent ought to think, not what one in fact does think. To claim that the norms of logic or our ability to grasp them reduce to the past success of reasoning according to these norms is an egregious example of the naturalistic fallacy, attempting to reduce the *ought* of prescription to the *is* of description. That Karl or his ancestors have done well by thinking logically does not explain why Karl should think this way or how he can see that he should.

But what does this have to do with epiphenomenalism? Why cannot the defender of evolutionary naturalism simply claim that interaction with the historical environment is what explains our reasoning patterns and leave the problem of justification to one side? This response has two problems. The first problem is that the prospects of a justification of rationality independent of natural selection but compatible with evolutionary naturalism look grim. Since logical truths, such as the truth that *modus ponens* is valid, have an eternal and necessary status, it is impossible to reduce them to facts about a finite, temporal, and contingent universe. Nor is there any hope of reducing logical necessity to nomological necessity since the same logical truths would hold in universes with different laws of nature. The most plausible justification of logical and mathematical truth is some form of Platonism. But Platonic objects are eternal and immutable: they are outside space and time and not subject to physical laws. Consequently, by any reasonable criteria, Platonic objects are nonnatural. What is more, that means that our ability to grasp logical truths depends on a nonnatural relation between our minds and Platonic objects. Thus, if the evolutionary naturalists try to help themselves to Platonic objects, they are abandoning naturalism.

The second problem with the evolutionary naturalist's response is that it overlooks an important distinction between two ways in which a creature can be rational. A good analogy is Kant's famous distinction between acting from duty and acting in accordance with duty. Kant pointed out that one can act in accordance with duty (do what duty requires) even though one has entirely nonmoral motives. In that case one does not act from duty (do one's duty because it is one's duty). Likewise, a creature's brain and nervous system can act in accordance with reason without the creature's acting from reason, that is, without its being the case that rational thought explains the creature's behavior. As we saw in the empirical argument for epiphenomenalism, rational patterns that tend to promote our survival can be implemented automatically and quite independent of thought. Therefore, if rational thought is not to be a redundant rider, it must be that rational thought is not automated in this way but is under the control of the agent's will. But if that is so, the agent must have the power of deliberation, and that means that one must be able to see whether a conclusion follows and thereby see whether the conclusion is what one ought to think or do. Since evolutionary naturalism has no plausible account of how such logical insight is possible, it is not able to justify a substantive role for rational thought. But if automated reason is doing all the work, evolutionary naturalism should predict that rational thought does not play a role in the production of behavior. Yet if that is correct, the contents of our thoughts can be organized randomly. What is more, they most likely would be, for two reasons: one, logical thought is tightly constrained, and there are many more irrational ways of organizing thoughts than rational ones; two, rational thought has nothing to contribute, so there is no reason why it would be favored. As long as our neurophysiology acts in accordance with reason, our thoughts can be and most likely would be a circus of irrationality. We would then act in accordance with reason yet not because of rational insight—that is, we would not act *from* reason. As a result, there would be no grounds for expecting that our beliefs are usually true or that our desires are usually appropriate.

Thus, there are strong reasons for saying that if evolutionary naturalism is true, epiphenomenalism is true. But if epiphenomenalism is true, there is good reason to expect that our cognitive mechanisms are unreliable. This counts against not only ST but also the weaker WT: evolutionary naturalism is incompatible with the reliability of our cognitive mechanisms in that it implies that these mechanisms are most likely unreliable.

Folk Psychology and Instrumentalism

Suppose, however, that evolutionary naturalism can overcome the threat of epiphenomenalism and that the content of intentional states does play a causal role in the production of behavior. Why suppose that A2 is true, that it is still unlikely that our cognitive mechanisms are reliable? In fine, because natural selection operates at the level of what is useful, not what is true or appropriate. Skinnerian creatures have a useful mechanism, one that allows operant conditioning so that a creature can adapt its behavior to novel environments within its own lifetime. Proto-Popperian creatures are somehow gifted with intentional states and thus the capacity to build mental models. Is it likely that proto-Popperian creatures will have mostly accurate representations? It is easy to assume that the answer is yes because, as Dennett says, when we adopt the intentional stance toward a creature, we assume that it will conform to norms of rationality. However, we are not entitled to make this assumption here. Since we are asking whether natural selection can explain the emergence of rationality, we are not entitled to assume that norms of rationality apply without adequate justification. On an evolutionary account, rationality is something that somehow develops from irrationality (or at least nonrationality): this transition is the one that the evolutionary naturalist needs to explain. As we have just seen, there are formidable obstacles in the way of this project since automated reason is a much more likely development of Skinnerian creatures than rational thought. So whatever it is that intentional content causally contributes, it provides no reason to think that those states are true or appropriate. Second, as Plantinga has argued, it is not hard to see how intentional states could be configured in such a way that they do cause adaptive behavior, even though they are mostly false or inappropriate:

> Perhaps a primitive tribe thinks that everything is really alive, or is a witch or a demon of some sort; and perhaps all or nearly all of their beliefs are of the form *this witch is* F or *that demon is* G: *this witch is good to eat*, or *that demon is likely to eat me if I give it a chance*. If they ascribe the right properties to the right

> witches, their beliefs could be adaptive while nonetheless . . . false. Also . . . it is belief and desire . . . that together produce behavior. But then clearly there could be many different systems of belief and desire that yield the same bit of adaptive behavior, and in many of those systems the belief components are largely false.[32]

Some critics are unconvinced by these scenarios. Jerry Fodor argues that even if a creature thinks of everything as a witch or a demon, the creature will still have many true beliefs if its behavior is adaptive:

> Let it be that *that appletree witch is blooming* is false, or lacks a truth value, in virtue of its presupposing that that appletree is a witch. Still, much of what a creature believes . . . is straightforwardly true. For example: *that's an appletree*; *that's blooming*; *that's there*; *something is blooming* . . . and so on indefinitely.[33]

Fodor's point is that a creature's radical misconception about the world does not entail that most of its beliefs are false. It needs further argument to show that a creature's mental model could be adaptive yet ubiquitously in error. A second challenge is made by William Ramsey. Suppose Plantinga can describe scenarios in which a creature's representations are adaptive yet mostly inaccurate. This still does not show that there is a plausible mechanism that systematically generates such representations and that may have been available for selection:

> Instead of giving us specific belief-desire pairs that produce life-preserving behavior in certain odd circumstances, Plantinga needs to describe general *mechanisms* or *processes* that would prove adaptive while giving rise to a preponderance of false beliefs.[34]

I think that both of these challenges are fair. To answer them effectively, one has to, first, provide a systematic means of generating false or inappropriate but nonetheless adaptive representations and, second, show that the tendency to misrepresentation will be so ubiquitous as to contaminate most of what a creature thinks. In my view there is a clear way of meeting these requirements, one that does not require that all of a creature's beliefs involve propositions of a particular logical form or any other gerrymandering.[35]

Consider another proto-Popperian creature, one that has intentional states whose content is causally relevant to the creature's behavior. Call this creature Pierre. It is not hard to see how Pierre could have adaptive responses superior to its ancestors even though its representations are mostly inaccurate. We need only exploit the parallel between theoretical instrumentalism and psychological instrumentalism. It is a matter of record that many scientific theories have been useful computational devices even though they are substantially false. Although Ptolemy's theory postulated nonexistent epicycles and equant points and although it was an almost entirely false model of the solar system, it still predicted the orbits of the known planets with considerable accuracy. In that sense, we might say that Ptolemy's model was both useful and "adapted" to the astronomical environment. If, like Popper, we extend the idea of natural selection to scientific theories, then Ptolemy's theory was selected because it was adapted to the observable phenomena, not because it was substantially true. The fact that observations are theory laden guarantees that fundamental misconceptions in theory will contaminate even the most basic observation statements so that error will indeed be ubiquitous. When the Ptolemaic astronomer observes the apparent retrograde motion of Mars, the astronomer falsely reports observing Mars rotating backward on its epicycle while the epicycle moves more slowly around a deferent, circling Earth. Not everything but almost everything the Ptolemaic astronomer says about planetary motion is false. The Ptolemaic astronomers can make a lot of mundane astronomical statements that have some truth in them, such as *That is a planet*, *Mars is a planet*, and so forth;[36] but they are swamped by errors about what planets are, the relation of Mars to other planets, the orbits of Mars and other planets, the axes of rotation, and on and on.

Likewise, selection will favor those proto-Popperian creatures whose mental models are useful computational devices—that is, devices that, when given environmental stimuli as input, compute adaptive responses as output. In and of itself, the truth of these mental models is irrelevant. All that matters is that these models are configured in some way that produces the right behavior. A precise analogy emerges between the predictions of a scientific theory, which keep it in business, and the behavior of a creature, which keeps it alive. Suppose that Pierre is an animist who thinks of everything that occurs as the favor, displeasure, or indifference of the spirits that inhabit everything around him. When lions attack, Pierre believes a spirit is displeased, so he runs to hide from its anger. When creatures and fruit are available for him to hunt and gather, Pierre believes that the spirits are happy with him, and he is eager to accept their gifts so as not to arouse their displeasure. When he is neither in danger nor in the presence of food or a mate, Pierre supposes that the spirits have other things on their mind and that he must seek out their blessing by looking for their gifts. Because Pierre is effective at finding and accepting food and at avoiding danger, Pierre is well adapted to his environment. His mental model consistently leads to adaptive behaviors. Nonetheless, the majority of Pierre's beliefs are false, and many of his desires are inappropriate. All his beliefs about where benefits and dangers come from are false, and his desires to please or propitiate the spirits are misplaced. Pierre's mental model is like the Ptolemaic astronomer's theoretical model: a relatively useful computational device that obtains the right output yet is nonetheless systematically inaccurate. This is enough to show that, given evolutionary naturalism, psychological instrumentalism is at least possibly true and, if so, that creatures could have adaptive behavior despite their having unreliable cognitive mechanisms.

However, it is not enough to point out that psychological instrumentalism is possibly true. Branden Fitelson and Elliott Sober would object, as they have to Plantinga, that my account "ignores the fact that the probability of a trait's evolving depends not just on its fitness, but on its *availability*."[37] Before we can claim that natural selection would favor creatures like Pierre, it must be shown that, assuming epiphenomenalism is false, the mental models available to creatures are more likely to have been psychologically instrumentalist than psychologically realist.

Unfortunately for evolutionary naturalism, it is easy to show this, using another analogy with theoretical instrumentalism. Suppose that we have exactly three data points (represented by crosses) all at the same distance from a horizontal axis. These three data points represent all that is actually observable of a given phenomenon. The correct theory is the one that shows how these data points are related, and it is represented by a particular line that connects all three points only once.

+ + +

Suppose further that in point of objective fact the correct theory is the aesthetically simplest one: the data points should simply be joined by a horizontal straight line. However, there are an infinite number of other ways of joining up these data points, corresponding to an infinite number of false theories that cover the same data. Because these data points represent all that we can observe of the phenomenon, these theories are observationally equivalent to the correct theory. So there is an array of instrumentally acceptable but false theories and only one true theory. Without relying on aesthetic tests, such as simplicity whose reliability is currently in question, it is therefore always more likely that an unbiased observer will select an instrumentally adequate but false theory.

All we have to do to transfer this to the psychological case is to think of the data points as representing stimuli in the environment and adaptive responses. Lines joining the data points represent mental models. Any correct mental model will consist entirely of beliefs and desires true or appropriate of these data points. Unlike our rather artificial theoretical case, many sets of beliefs and desires may be able to fit the bill; even some suboptimal collections would be reliable enough to be acceptable since our definition of cognitive reliability only requires that most of a creature's intentional states are true or appropriate. Believing that a growling creature is dangerous may be just as good as believing that that lion is dangerous. Desiring to flee is perhaps no better than desiring to be somewhere else. However, given the finite number of data points (stimuli and responses) and supposing that these exhaust the stimuli that a creature can observe and the responses it can produce, the basic lesson of theoretical instrumentalism remains: There are vastly more false mental models that cover these data points than there are accurate or nearly accurate ones. Therefore, it is much more likely that psychologically instrumentalist models will be available for selection than psychologically realist ones. Furthermore, since these false mental models correlate the stimuli with adaptive responses just as well as the true ones, they are indistinguishable to natural selection. Thus, there is no reason to expect that psychological realists would evolve from creatures whose ancestors were psychological instrumentalists. Hence, given evolutionary naturalism, it is far more likely that psychological instrumentalism is true than that psychological realism is true. But in that case, it is most likely that although our behaviors are adaptive, our cognitive mechanisms are unreliable. This counts against not only ST but also the weaker WT.

However, if our cognitive mechanisms are unreliable, then it is not possible to rely on our scientific theories as guides to the truth about the world and what belongs to our ontology. Indeed, if psychological instrumentalism is true, then surely theoretical instrumentalism is true as well. Evolutionary naturalism therefore cannot be relied on to state the truth about how our minds—or anything else—originated. Rather, it is at best a useful calculation device consistent with observed phenomena.[38] Scientism,[39] which relies on evolutionary naturalism to dictate our ontology, cannot be sustained.

What Else Could It Be?

At this point the default response of the evolutionary naturalists is invariably "Yes, these are all problems, but what else could it be?" So long as they think there are no credible alternatives, the evolutionary naturalists treat the most powerful objections to their view as "difficulties in theory" that will at some point be addressed or that are just beyond the ken of "midrange primates." Why not rearrange the deck chairs on the Titanic if there is no way to jump ship?

However, evolutionary naturalism does have a compelling alternative. While not the only logical possibility, theism is the most fully developed expression of this alternative. According to theism, reason is prior to nature. God Himself is supremely rational, and our rationality is not something that evolved from the nonrational but is rather an imperfect participation in divine rationality. Since theists reject naturalism, they have no difficulty in endorsing Platonic accounts of logical and mathematical truth. Logical and mathematical insight is then no longer fallaciously assimilated to the past experience of a creature or its ancestors. Rather, such insight is precisely what it appears to be, a genuine grasp of nonnatural, eternal, and necessary truths.

As Robert Koons has shown, this sort of approach can be extended to the inductive case and our aesthetic criteria of theory choice.[40] Granted evolutionary naturalism, there is no way to justify the idea that simplicity, symmetry, and beauty are indicators of truth. All we can say is that these have been useful instruments, ones that have served us well up to now but that may be in error. For the theist, however, there is independent reason to trust certain criteria of theory choice. Since God is a rational being and since simplicity, symmetry, and beauty are marks or signs of rationality, we have reason to expect that creation actually exemplifies these marks. It would be arrogant to suppose that the precise form of these marks is something that our limited minds can always anticipate or discern. Nonetheless, in the theistic scientist's humble attempts to think God's thoughts after Him,[41] if such signs manifest themselves, the scientist has good reason to expect they are pointers to truth, not merely the imposition of psychological categories that happen to have been adaptive.

Theism also gives a better general explanation of the reliability of our cognitive mechanisms. Given theism (at least of the Judeo-Christian variety), psychological realism is a plausible thesis since we are made in the image of a rational God who knows all truth. Of course, we do have mental frailties—which evolutionary naturalists are quick to point out—and that reminds us that we have imperfect, fallible minds. These are minds with a basically sound design but in a somewhat degenerated condition, minds that are basically reliable yet prone to bias and suboptimal in functioning. Thus, fallible psychological realism—rather similar to Karl Popper's falliblist realism, about science yet with a bit more confidence in induction—is the prediction of theism. As a result, theists can sustain the rationality of science and its authority to tell us, at least provisionally, what sorts of things belong in our ontology. By contrast, as we have seen, the best evolutionary naturalism can justify is an emaciated psychological instrumentalism. If evolutionary naturalism is true, it is likely that we do not and cannot know what is really going on under the veil of the "manifest image" of appearances; and as a result, science has no authority to tell us what to believe. Sooner than give up on the rationality of science and its ability to discover truth, evolutionary naturalists should jump ship to theism or at the least to a less theologically committed version of intelligent design.

Again, the objection will be that the problem with theism or design is that although it would solve lots of problems if it were true, there is no way it could be true. For example, one argument is that because causation is operative only in space and time, there cannot be a causal mechanism linking the Platonic realm and the natural order. Thus, although access to Platonic objects would explain our logical and mathematical insight, it cannot be the correct explanation, since such objects could not interact with our minds. In response, it should first be noted that this objection is question begging in that it assumes that the relation between a nonnatural entity and a natural one must be naturalistic causation. But why on earth should we suppose that to be the case? Why not suppose that there is *trans*natural causation connecting the nonnatural and the natural? In fact, as Koons has argued, the reliability of simplicity as a guide to truth requires such an explanation:

> By definition, the laws and fundamental structure of nature pervade nature. Anything that causes these laws to be simple, anything that imposes a consistent aesthetic upon them, must be supernatural.[42]

Koons's point is that if nature has a completely global property—such as the simplicity of its laws and fine structure constants, a property that permeates everything in space and time—then this property clearly cannot be explained by anything in space and time or even by space and time in its entirety. For any such explanation would invoke a contingent entity that has the very property we are trying to explain, and thus it would be appropriate to ask from whence that property derived. The original problem, of why any such property is exemplified at all, would merely be displaced. The only way to avoid this dead end is to infer a supernatural cause, some being that is able to confer the property on all of nature. Regress can be avoided by supposing that the source of this property is some eternal, necessary being whose reason exemplifies that property essentially. Since such a being is not contingent, no further request for explanation of the property's origin is warranted or even intelligible. Now if the only credible explanation of something is supernatural (or nonnatural), then we have good evidence that transnatural causation is possible.[43] If science is really open to following the evidence wherever it leads, then that means abandoning evolutionary naturalism and scientific materialism.

Conclusion

In this chapter I have defended the case advanced by C. S. Lewis and Alvin Plantinga for thinking that evolutionary naturalism is incompatible with the reliability of our cognitive mechanisms and is thus (epistemically) self-defeating. If evolutionary naturalism is true, then it is most likely that epiphenomenalism is true, in which case our beliefs and desires can widely diverge from reality. But even if epiphenomenalism is false and selection could find a role for intentional contents, psychological instrumentalism is far more likely true than psychological realism. If so, there is still no reason to think that our cognitive mechanisms or our scientific theories are reliable. Those who cherish the rationality of science and its authority to pronounce on likely ontology must either retreat to a muted instrumentalism or abandon evolutionary naturalism and scientific materialism.

Notes

A version of this chapter appeared in *Philosophia Christi* 5, no. 1: 143–165. Thanks to Ed Martin of Liberty University and Trinity College for his helpful comments on an earlier version of this chapter.

1. C. S. Lewis, *Miracles*, 2d ed. (New York: Macmillan, 1960), 21.
2. Lewis, *Miracles*, 105.
3. Genetic drift and other undirected processes of evolution that are not strictly Darwinian can also be included.
4. While the focus of this book is strictly scientific materialism and it is perhaps conceivable that someone is a scientific materialist without being an evolutionary naturalist, evolutionary naturalism is at the least a strong commitment of current scientific materialists, the refutation of which would leave them in considerable disarray.
5. Lewis makes this case in chapters 3, 4, and 13 of *Miracles*.
6. Alvin Plantinga makes this argument in "Is Naturalism Irrational?" chapter 12 of his *Warrant and Proper Function* (New York: Oxford University Press, 1993). A later version of the same argument, including a technical correction and some helpful simplifications, is presented in Plantinga's *Warranted Christian Belief* (New York: Oxford University Press, 2000). Plantinga has responded at length to his critics in "Reply to Beilby's Cohorts" in *Naturalism Defeated: Essays on Plantinga's Evolutionary Argument against Naturalism*, ed. James Beilby (Ithaca, N.Y.: Cornell University Press, 2002).
7. I use the word "justification," not "warrant," since we are talking about the absence of a necessary condition for rational belief, not knowledge. Thanks to Ed Martin's advice on this point.
8. Plantinga distinguishes "epiphenomenalism simpliciter," according to which beliefs have no causal role, from "semantic epiphenomenalism," which allows that beliefs might have some causally efficacious properties but denies that their content

is one of them. See Plantinga's *Warranted Christian Belief*, 231–36. For the purposes of our argument, however, this distinction is not important, because on both versions of epiphenomenalism, the contents, and therefore the semantic logical properties of belief, have no connection to the natural causal nexus.

9. Plantinga, *Warranted Christian Belief*, 236.

10. Plantinga, *Warranted Christian Belief*, 235.

11. Although Lewis never quite made this argument, it is strongly suggested by some of his remarks in chapter 3 of *Miracles*, including the first quote at the head of this chapter. My argument actually makes a stronger claim than Plantinga's. Plantinga thinks that if content has a causal role in producing behavior, then the probability that our cognitive mechanisms are reliable is, at best, moderately high. I argue that it is still low.

12. Of course, we frequently rely for action on information that is not reliable in the sense I am discussing. Even if an informant lies 80 percent of the time, it may be prudent to clear a building when that person claims to have learned that it will be bombed. In that case we "rely" on what in our sense is an unreliable source of information. Note, however, that we are not rationally required to believe the informant: we are just acting on the assumption that there is a significant chance that the person may be telling the truth; and given the possibly dreadful consequences of not acting on that assumption, we act accordingly. Likewise, suppose Amanda learns that she has a one in ten chance of winning a million dollars if she completes an application. She "relies" (acts on) this chance because she has nothing to lose and much to gain by trying, but it does not follow that she should believe that she will win. As we understand it, a cognitive mechanism is reliable only if it can be depended on for truth, not merely for action.

13. For example, this would not account for the good of altruistic actions, a major area of controversy between sociobiologists and their critics.

14. This condition is similar to the one proposed by Plantinga, who assumes that cognitive reliability means that "our cognitive faculties are reliable . . . in the sense that they produce mostly true beliefs in the sort of environments that are normal for them" (*Warrant and Proper Function*, 220). One could say a vast amount about normal conditions or environments, but the basic idea is that the unreliability of a cognitive mechanism in an odd case does not show it is unreliable in normal cases. For example, the fact that an expert at critical thinking might be brainwashed while under the influence of certain drugs hardly shows that the person's belief-forming mechanism is unreliable in the normal case, where no such drugs are at work.

15. W. V. O. Quine, "Natural Kinds," in *Ontological Relativity and Other Essays* (New York: Columbia University Press, 1969), 126.

16. Fodor, "Is Science Biologically Possible?" *In Critical Condition*, 197.

17. Dennett, "True Believers," *The Intentional Stance*, 33.

18. Fodor, "Darwin among the Modules," *The Mind Doesn't Work That Way*, 88.

19. Fodor, "Is Science Biologically Possible?" *In Critical Condition*, 201. Fodor uses "true believer" as shorthand for a creature with reliable cognitive mechanisms.

20. To be fair, perhaps Fodor thinks there is some other explanation of psychological realism than natural selection, and some of his remarks even suggest that this is his view. However, if this is right, Fodor owes us as at least an outline of how this alternative account goes.

21. Here I will understand epiphenomenalism in the broad sense as claiming either that intentional states do not cause behavior at all (classical epiphenomenalism) or that although intentional states do cause behavior, the intentional content of these states is nonetheless causally inert or irrelevant (semantic epiphenomenalism). In both cases, what we believe and desire would have nothing to do with our behavior.

22. I do not wish to be dogmatic here. Perhaps Fodor thinks that there is a naturalistic way of getting at reliability independent of selection, but if so, it is unclear to me what that is.

23. Dennett, *Darwin's Dangerous Idea*, 373–83.

24. Dennett, *Darwin's Dangerous Idea*, 374–75.

25. Dennett, *Darwin's Dangerous Idea*, 375.

26. See especially chapter 3, "The Cardinal Difficulty of Naturalism," in Lewis's *Miracles*.

27. Lewis, *Miracles*, 19. The only caveat I would make is that Lewis overlooks an ambiguity in "rational." Calculators, computers, and robots are highly rational in the sense that they implement rational principles, but they are not rational in the sense that their own logical thinking is the explanation of their output.

28. A good example of such a metaheuristic is the so-called killer heuristic used by AI ticktacktoe programs: "the computer gives precedence to investigating opponent responses that killed, or refuted, other moves the computer has already considered. [Suppose] the killer move was for crosses to play center. . . . The computer would look for the opponent's move first. . . . It would then discover that this play also leads to a weak position for [the computer], and immediately assign a weak rating to the entire branch" (Daniel Crevier, *AI: The Tumultuous History of the Search for Artificial Intelligence* [New York: Basic Books, 1993], 229). Since the "entire branch" is essentially a game-playing strategy or heuristic, the "killer heuristic" is a metaheuristic for adjusting lower-level heuristics.

29. Lewis, *Miracles*, 20.

30. It seems to me that evolutionary naturalism implies that the problem of induction is insoluble; therefore, the faith of a scientist, in some objective order for him or her to discover, is irrational. Once again, scientific materialism undercuts the rationality of science.

31. Lewis, *Miracles*, 20.

32. Plantinga, *Warranted Christian Belief*, 234–35.

33. Fodor, "Is Science Biologically Possible?" *In Critical Condition*, 193.

34. William Ramsey, "Naturalism Defended," in *Naturalism Defeated?* ed. James Beilby, 20.

35. The problem with gerrymandering is that it only enables us to show that there is a perhaps quite "close," possible world in which evolutionary naturalism holds; but

thought is unreliable. It does not convince us that that is the way our world most likely is if evolutionary naturalism holds, since it may just be a fact that the gerrymandered conditions did not obtain. The evolutionary naturalist is quite happy to admit that the reliability of our cognitive mechanism is a contingent fact and that, had circumstances been different, those mechanisms would not have been reliable. It is therefore necessary to confine our thinking to what was most likely the case in our world, granted naturalistic assumptions.

36. Some truth but not as much as one might think. The Ptolemaic astronomer defines "planet" as "wanderer around the Earth," wrongly includes the sun and moon in the set of "planets," and wrongly excludes the Earth from that set. The Ptolemaic conception of the planets is neither extensionally nor intensionally equivalent to that of modern astronomy. The Ptolemaic astronomer is only right to say that *Mars is a planet* in the sense that Mars appears to be a wanderer about the Earth.

37. Branden Fitelson and Elliott Sober, "Plantinga's Probability Arguments against Evolutionary Naturalism," *Pacific Philosophical Quarterly* 79 (1998): 115–29, 120.

38. It is not consistent with observed phenomena, as I argued in chapters 4 and 5. Nonetheless, given our argument, observational consistency is the most that the evolutionary naturalist could claim.

39. Jerry Fodor is an exception. He defends scientism but resists the idea that evolutionary accounts are our best guides to reality, especially psychological reality.

40. Robert Koons, "The Incompatibility of Naturalism and Scientific Realism," in *Naturalism: A Critical Analysis*, ed. William Lane Craig and J. P. Moreland (London: RKP, 2000), 49–63.

41. This is the attitude to science of the famous Lutheran astronomer Johannes Kepler.

42. Koons, "The Incompatibility of Naturalism and Scientific Realism," 55.

43. The only alternatives here are either skepticism or instrumentalism, but they both mean giving up on any robust notion of the rationality of science.

CHAPTER SEVEN

~

Intentionality, Information, and Displacement: The Legitimacy of Design

> Perhaps . . . the intentionality of our cognitive attitudes . . . , a feature that some philosophers take to be distinctive of the mental, is a manifestation of their underlying information-theoretic structure.[1]

> There is no getting around the displacement problem. Any output of specified complexity requires a prior input of specified complexity. . . . There is only one known generator of specified complexity, and that is intelligence.[2]

At the end of chapter 2, I consider the kind of eliminativist skeptic who believes that design is a fictional category, a concept without a Kantian deduction. If that is so, design is like fate, a notion devoid of application; and chance and necessity are left as the only legitimate categories in science. I urge that this view is mistaken for two reasons. First, the very existence of concepts points to design because concepts are intentional and intentionality is what makes design possible. Second, even if the existence of concepts is denied and human cognition and action is explained in terms of the flow of information through a physical system, the kind of complex specified information involved in theoretical and practical reason still implicates design. It is now time to provide some serious arguments to develop and defend these intuitions in more detail.

One might think that the way to show that design is real is to show that only design accounts for some observable phenomena in the real world.

Thus, in chapter 4, I argue that top-down design seems essential to explain the appearance of functional unity and cohesion exhibited by irreducibly complex biological structures. However, the skeptic can claim that I have only given evidence that design is empirically useful, which is consistent with its being a fiction, akin to epicycles, equant points, or phlogiston. Indeed, the skeptic might go further and claim that design is not really required because observed entities only appear to be irreducibly complex or bear complex specified information. In the face of such a challenge, empirical arguments cannot help, based as they are on the appearance of complexity. For no matter how good such arguments might be, they are still compatible with the complexity's being fictional.

Nonetheless, threads have already appeared in this book that point to a different kind of argument. Such an argument would not depend on empirical observation of the external world but on an analysis of the character of human thought. In chapter 3, I argue that the intentionality of thought seems to be irreducible to undirected causes. And in chapter 5, I demonstrate that design is crucial to understanding the structure of our theoretical and practical reasoning. In this chapter, I extend these lines of reasoning to argue in more detail that the legitimacy of design is established by the character of thought itself. To the skeptic who claims that design is a mental construct imposed on external reality, I reply that those constructs themselves exemplify the marks of design. In particular, it is precisely those mental structures that we call "plans" or "designs" that evince actual design. But if design is a real feature of thought, then we should be open to the possibility that design is also exemplified by the external world. And indeed, this is demonstrably the case. For, as I will show, human plans for actions are connected to actions in such a way that if the former are designed, so are the latter. But if human agency results in undeniably designed effects, it is dogmatic to exclude the possibility that other kinds of agencies are capable of doing the same thing.

I start with the more informal and intuitive of my two arguments, the argument from intentionality. This argument aims to show that, granted the existence of concepts, intentionality and design cannot be fictional concepts since their legitimacy is established by the very nature of concepts and their connections to the world. Next, I address the response of those who claim that the very idea of a concept is an antiquated and redundant notion. According to this view, cognition and action can be fully accounted for in terms of the passage of information through physical systems. I offer two main responses. First, I note that for information to do the explanatory work we want, intentionality is still required; then, I give a more formal argument that requires a short digression into philosophical information theory. With this foundation, the legitimacy of design is established via an application of the recent work of William Dembski to the complexity of human cognition and action.

The Argument from Intentionality

Before one can use the fact of intentionality as an argument against the causal sufficiency of undirected causes, it is necessary to show that intentionality is real. If intentionality is merely an appearance, it could be just an illusory by-product of the forces of natural selection operating on our brain. But as I have pointed out on several occasions, it is not possible for intentionality to be merely an appearance. Here I will state the argument for this thesis with a bit more rigor.

First, I adopt a position of hyperbolic doubt, not about our sense experiences, but about our concepts. That is,

> A1: Suppose that we are subject to cognitive illusion as much as possible, with the exception that our faculties of logical analysis are reliable.

By "cognitive illusion," I mean that our concepts are fictional, like fate or fortune, and do not carve nature at the joints. However, if we are to argue at all, we must suppose that this cognitive illusion does not extend to the concepts of logic themselves. Otherwise, we cannot so much as see what follows from what, nor can we detect contradictions and the like. This is surely a fair exclusion from cognitive illusion because our opponent, the naturalist, must also rely on logic. By our definition:

> A2: A cognitive illusion is (or involves) a fictional concept.

We now also need to define intentionality.

> A3: A mental entity *E* is intentional if and only if *E* has the following properties:
> 1. *E* can characterize nonexistent objects, for example, one can believe that Santa came down the chimney, even though there is no Santa.
> 2. *E* can characterize an object as being *F* and not *G*, even if all and only *F*'s are *G*'s and even if this is true as a matter of physical law or logical necessity; for example, one may believe a gas has been heated without believing that the MKE of its molecules has increased, or one may believe that there are seven deadly sins without believing that there are the prime-number-succeeding-the-number-five deadly sins.

Next we observe that

A4: Concepts are intentional.

This should be obvious, but we can easily prove it. One can conceive of the largest integer, even though there is no such thing; one can conceive that a gas is heated up, without conceiving of a change in its molecular motion; and one can conceive that there are ten commandments without conceiving that there are the cube root of one thousand commandments.

However, if concepts themselves are intentional, then no matter how fictitious their content may be, intentionality must really exist. Confuse my thinking as much as you may, you will never make me doubt the existence of intentionality, since to be confused is to have fictional concepts that are intentional mental entities. So, we conclude:

C: Intentionality is real.

The only way to avoid this conclusion is to deny that concepts really exist, a move that we will address in later sections of this chapter. So long as the eliminativists grant the existence of concepts, however, they cannot simply offer an account of why there appears to be such a thing as intentionality, although there really is not. They must explain the fact of intentionality yet do so without invoking design.

However, several reasons exist for thinking that this project cannot succeed. The most obvious of these is that if intentionality is a real characteristic of human cognition, then there is significant reason to think that design is exemplified by reality. This is so because among the things one can conceive of is a future project; some plan of action; or an idea for a book, proof strategy, or new experiment. But such conceptions are precisely designs, and the empirical fact that human actions are correlated with these designs provides excellent and abundant evidence for the existence of designed objects. The concepts provide a specification that is realized in the actions themselves. To claim that the concepts are epiphenomenal is extremely implausible because actions track conceptual specifications in systematic and predictable ways. So not only does the existence of human preconceptions establish the reality of design, but human action also gives us strong evidence that design is a causally efficacious category and therefore something which cannot be ignored in science. Those naturalists who insist that the only legitimate scientific categories are causal ones cannot use that principle to exclude design.

A second argument proceeds by examining the resources available to the eliminative naturalist. These resources—chance, necessity, or a combination of the two—are all varieties of undirected causation. But the most important feature of our concepts is that they provide directed causation. The plan for an experiment precedes the experiment itself and guides the experimenter's actions. There is, to be sure, no logical contradiction in the idea that undirected causation could throw up the capacity for directed causation. As Hume taught us, the conceptual character of an effect need not be deducible from the conceptual character of the cause. But there are, all the same, good reasons for thinking that undirected causes would not produce the capacity for directed causation.

The most important general argument to this effect is that the undirected causal nexus is not governed by norms of rationality or, if it is in some sense, not by the kind of norms that constrain the attribution and interpretation of intentional states. The key point is not that mental states are "holistic," in the sense that the appropriate state to attribute to an agent depends not merely on one's behavior but also on one's other beliefs and desires.[3] Holism alone does not get us very far, because, as Crane and Mellor point out, purely physical systems are also holistic:

> Newtonian force (*f*) and mass (*m*) are also conceptually interdependent, being partly defined by the relation $f = ma$. . . . And this relation too requires forces and masses to combine to produce their effects (accelerations)—and lets many combinations cause the same effect. So we can no more infer a force *f* or a mass *m* from the acceleration *a* they cause than we can infer a belief or a desire from the action they cause.[4]

It is not holism per se but the fact that the holistic attribution depends on the rational coherence of the set of intentional states that argues against such states arising from undirected causes. Crane and Mellor themselves seem to deny this position because, as they point out, the application of Newton's laws to phenomena is also constrained by rationality: the overall distribution of forces, masses, and accelerations must "make sense" in the light of Newton's laws.

The point is unconvincing, however, because laws and descriptions of observations are statements and are thus bearers of intentional content. That they must cohere is a result of the fact that statements in general are constrained by logical norms. But logic applies to thoughts and to contents, not to the causal processes themselves. The fact that scientific prediction and disconfirmation are governed by *modus ponens* and *modus tollens* is not a good reason for thinking that causal processes themselves implement these logical principles. Nor can one really infer that masses, forces, and accelerations are logical in the way that the theory describing them is because the masses, forces, and accelerations do not have contents and hence do not imply or give reasons for anything. The natural "because" in "e2 happened because e1 happened" is not the same as the logical "because" in "we must attribute acceleration a^* to the second particle because we attributed acceleration a and mass m to the first particle which struck it." Only the second "because" is a connection between reasons, and it is reasons and not undirected causes that have contents. The reason this point is obscured, I think, is because of a failure to make a distinction, noted in chapter 6, between acting in accordance with reason and acting from reason. Systems of undirected physical causes do indeed act in accordance with reason, which is why they can be described, explained, and predicted in logically coherent ways. But that is no reason to think that the undirected causes are governed by norms of rationality. These norms apply to agents, those who act from reason; they also apply to the intentional products of these agents, including scientific theories and explanations. But undirected causes are neither agents nor the intentional products of agents. Thus, while Newton's theory, being the intentional product of an agent, is governed by norms of rationality, it does not follow that the undirected causes to which the theory applies are also so governed.

Indeed the only plausible way for a naturalist to bridge the gap between undirected causes and directed causes is by providing a naturalistic theory of content. However, as we saw in chapter 3, the prospects for such theories are dim because causal theories of content cannot work without being relativized to the norms of proper functioning of an organism's cognitive architecture. While these norms are assumed to account for the character of beliefs and desires, the origin of the norms is never explained. But if norms of rationality need to be explained, it is scarcely an advance to offer unexplained norms of proper functioning. Perhaps this does offer some insight into the nature of rationality, but it clearly does not show how the normative could derive from the nonnormative. As we saw in chapter 3, it won't help to invoke Mother Nature. Either this means natural selection, a process devoid of norms and one that is very unlikely to produce norm-governed intentional states; or it means a hagiographical natural agent, in which case intentionality is being implausibly relocated but not explained. The problem with even the most promising naturalistic accounts of intentionality is that they seem only capable of displacing norms, not of showing how such things could arise from the nonnormative in the first place.

In some ways, this should not be surprising. Norms are prescriptive, characterizing what ought to be the case. But it seems that the very idea of prescription presupposes agents. If in fact we are supposed to believe and desire certain things in certain circumstances, the question "Supposed by whom?" is hard to avoid. When we try to say that because of natural selection, we are supposed to have these intentional states, we are falling back on a metaphor that treats natural selection as an agent. If we take the metaphor literally, we would have to ask why natural selection should have these intentions for us, which would require some final origin of intentionality. But either this leads to a regress or to some necessary being whose intentionality and agency requires no further explanation. However, if we do not take natural selection's intentionality literally, then we cannot explain the literal attribution of intentionality to human beings. Yet, if I am right, the correctness of such attribution is rock solid. Either way, the naturalist is unable to explain the emergence of norm-governed states.

The existence of contingent intentionality in human beings surely demands an explanation, but it does not seem that natural causes suffice to explain it. Because it is governed by norms of rationality, intentionality appears to be sui generis. Indeed Brentano was right to maintain that the intentional cannot arise from the nonintentional, and that this is a decisive objection to materialism. Consequently, the only coherent explanation of contingent intentionality is the existence of some necessary being, an agent from whom all other intentionality derives but who does not require further explanation.

This argument has force as philosophy, I believe, but is too informal to carry the day as a scientific argument. Perhaps, after all, the notion of a "concept" is antiquated, and we can say everything we need to say about the mind and its relation to the world in terms of information-bearing vehicles of some kind. In the quote at the beginning of this chapter, Dretske suggests that intentionality might not be the distinctive mark of the mental. Perhaps it is simply inherited from the natural structure of information-bearing signals and states. In that case, it seems that Churchland could evade much of my critique in chapter 2 by noting that all my objections depend on the existence of concepts and by suggesting that neuroscience only needs to speak of the passage of information through a physical system.

Against this suggestion, I have two lines of response. The first is to note that the move does not really work but is yet another example of relocating (and hiding) the original problem. If the transmission of information is to do any work in explaining human cognition and behavior, this information cannot be viewed as mere uninterpreted signals. We must suppose that the information has content. For example, if certain signals from the visual cortex of a tourist near Banff National Park carry the information that a grizzly bear is ten feet away, then that fact clearly explains the tourist's recollecting the ranger's advice, "Do not run; if attacked, lie face down until the bear moves away." But physical signals are not self-interpreting. Indeed the pattern of events occurring in the visual cortex might be interpreted in an infinite number of ways, only a few of which are relevant to surviving an encounter with a grizzly. The fact is that the salient information in these signals is only recoverable by an interpretive agent who understands them. Understanding is, however, an intentional state.

Likewise, merely understanding the information that a grizzly is nearby explains nothing without supposing that one not only has a desire to stay alive but also has instrumental beliefs about how to do so. It is not just information but certain attitudes toward information that we need to attribute to agents if we are to explain the way that they behave. A suicidal person or one with an inflated sense of his or her fighting prowess might produce different behaviors in response to the same information. So an information–theoretic account of cognition is not really an alternative to intentional psychology. For information to do the kind of work that it needs to do to explain human actions, intentional attitudes toward that information will have to be involved.

A second and more scientific response does not rest its case on the informal notions of concepts, attitudes, and intentionality. Instead, it calls attention to the information–theoretic complexity of human cognition and action. This complexity points to design rather than undirected causes. I develop this line of argument in chapter 5, in the section titled Reason, Practical and Theoretical. I point out that in an agent's practical or theoretical reasoning, the collection of reasons is often irreducibly complex, consisting of a number of tightly specified, well-matched, and interacting reasons, of which the removal of any one would prevent the occurrence of the concluding action or thought. In information–theoretic terms, one could say that an irreducibly complex nexus of reasons constitutes complex specified information, and this is very unlikely to be the product of undirected causes. Rather it seems best explained by the intelligent agency of the individual who has those reasons. It is not that the right kind of information simply happened to accumulate in the individual's brain but that his or her own agency was decisive in selecting and combining the relevant, well-matched reasons to issue in a particular action or thought.

The argument has force, I believe, although with a vagueness about just how complex an agent's reasons have to be before undirected causes can be eliminated as plausible explanations. As a recent critique of Behe pointed out, for irreducible complexity to be a plausible indicator of design, we need to assume that the system is "very complex."[5] Without specifying a rigorous, quantitative definition of "complex," it might be that some irreducibly complex systems are nonetheless not complex enough to conclusively exclude undirected causes. At this point, Dembski's work on complex specified information is helpful because it gives an explicit measure of the degree of complexity a system needs to exhibit before we can infer design. For the next two sections, I unpack these ideas and apply them to cognition and action explanation, which makes it possible to give a rigorous and, I believe, decisive argument in favor of the legitimacy of design. Wherever else it may retreat under fire, design cannot be eliminated as a legitimate category in explaining human cognition and action. But if so, the possibility of design beyond the human should be an open and scientific question, not one that is sealed off forever by aprioristic methodological strictures. This argument requires a short digression into philosophical information theory.

Philosophical Information Theory

As a science, information theory has been concerned with the average amount of information that signals carry when passing down a communication channel: "All the interesting theorems in communication theory depend on 'information' being understood as *average information*."[6] By contrast, philosophical information theory is interested in comparing the amount of information carried by individual signals. One important factor in conveying information is how many possibilities are eliminated. Tautologies, such as "This object either weighs ten pounds or it does not," convey no information because no possibilities are excluded. Every object in the universe is such that it either weighs ten pounds or it does not. Likewise, vague statements typically provide less information than precise ones. "Amanda did okay on the final" conveys very little information because "okay" is unclear. Did she pass the test, or did she flunk it but not so badly that her other tests and coursework cannot make up for it? "Amanda passed the final" conveys more information because it definitely rules out the possibility that Amanda failed. And "Amanda received a ninety-two" conveys even more information since, on reasonable background assumptions, it not only conveys the information that Amanda did not fail but also excludes the one hundred other scores she might have received.[7] Thus Dretske suggests that as a general rule "The amount of information associated with a state of affairs has to do, simply, with the extent to which that state of affairs constitutes a reduction in the number of possibilities."[8]

The account is not adequate, however. As Dembski points out, it "tells us nothing about how those possibilities were individuated."[9] Suppose we are playing poker, and we want to know how much information is conveyed when someone lays down a royal flush. If we only compare this possibility with its opposite, namely, not laying down a royal flush, then we eliminate only one possibility and consequently learn very little; yet, this scenario fails to account for the fact that it is typically surprising when someone has a royal flush. The surprise derives from the fact that the royal flush is far less probable than its opposite.[10] Roughly speaking, the amount of information we learn from a royal flush is inversely proportional to the probability, that is, if we assume that no "side information" has excluded certain possibilities. In other words, the more improbable an event, the more information its occurrence conveys.

Using the resources of information theory can make this concept more precise. Information theory has focused on the context of digitally encoded information transmitted between computers, so it is standard to measure information in *bits* (*b*inary dig*its*). In this way we can define the information content of a signal as the negative base-2 logarithm of its probability. Thus, if $I(s)$ is the information carried by s and $P(s)$ is the probability of s, then $I(s) = -\log_2 P(s)$. Now suppose that r represents someone's being dealt a royal flush, an event with a very low probability. The unconditional probability $P(r) = 0.000002$ and $I(r) = -\log_2 P(r) = 19$ bits, a very *large* amount of information.[11]

One can then represent the unconditional information I, generated by two events A and B as simply

$$\text{UI: } I(A \text{ and } B) = I(A) + I(B).$$

However, UI only works for probabilistically independent events. For example, what is the probability that two people selected as game show contestants by a random process have the same surname? If there is a probabilistic dependence between events, UI won't work. Suppose that the process for selection is only random with respect to families so that if any one family member is selected, then another family member will be selected as well, if there is one. Then the only times it could fail that two people with the same surname were chosen would be in cases where there were no other surviving family members; where another family's name had been retained as a surname despite marriage; or where the name had been changed by personal preference, remarriage, and so forth. Because of the probabilistic dependence between drawing one person with a surname and drawing another with the same surname, the overall probability of drawing two people with the same surname is higher than if the name drawings were independent and thus the information associated with that event lower. It might be very unlikely that anyone called Parker is selected, so one would gain a lot of information by learning that one such person is on the game show. But, given this selection, it is highly likely that another Parker would be selected as well, so it would give us very little additional information to learn that another Parker would be on the show. As a result, UI would overestimate the information gained by learning that two Parkers were selected for the game show.

This kind of case is better represented by the idea of conditional information. Let $I\ (B \mid A)$ denote the information generated by B given A. Then

$$\text{CI: } I(\text{A and } B) = I(\text{A}) + I\ (B \mid \text{A}).$$ [12]

According to CI, if one learns that any Parker was selected for the game show, this is still unlikely and so conveys a lot of information. But, given the rules for selecting contestants, very little additional information is gained by learning that another Parker was selected as well, since, given the selection of the first Parker, this was highly probable.

This suffices to explain how it is possible to talk about the amounts of information contained in a system. It is this understanding of measuring information that we will employ to show that human cognition and action are frequently too complex and tightly specified to reduce to undirected causes.

The Argument from the Complexity of Thought

Before we can state this argument, we need some principle that can tell us the limits, if any, to the informational resources available to the scientific materialist. According to scientific materialism, any effect, including the information content of any representation, can be explained by chance, necessity, or the combination of the two. It might seem that anything at all can be accounted for by such nebulous categories: either something had to happen, according to a deterministic natural law, or it did not, in which case it could be a chance event.

Unfortunately for scientific materialism, as Dembski has shown, chance and necessity have limits to what they can explain. In particular, Dembski argues that these categories do not suffice to account for complex specified information (CSI). Using Dembski's universal probability bound of 1 in 10^{150}, we can define an event as complex if its probability is less than 1 in 10^{150}. In information-theoretic terms, such an event generates an enormous 500 bits of information, signifying the vast number of alternative possibilities that its occurrence has eliminated. As unlikely as complex events are, they still could happen by chance or via the interplay of chance and necessity, that is, if they exhibit no intelligible pattern. However, Dembski has proposed that we can infer design if a complex event also conforms to a "detachable" specification. The details of detachability are quite technical, but the basic idea is clear. A detachable specification is a pattern that is not arbitrarily imposed after the fact—for example, drawing a target around an arrow embedded in a wall and declaring it a hit. It "cannot simply be read off the events whose design is in question,"[13] as when the Greeks claimed to "see" all kinds of constellations, which are in many ways arbitrary groupings of stars. Rather, a detachable specification is one that is, in a precise sense, independent of the event specified.

To use one of Dembski's best-known examples, suppose that scientists who are engaged in the search for extraterrestrial intelligence (SETI) receive a digital radio signal—and that it is complex, containing more than 500 bits of information. Also suppose that the scientists discover that this sequence is the unary coding for the prime numbers between two and eighty-nine, exactly in sequence.[14] That prime numbers exist and that they exhibit this sequence are facts we know quite independently of our learning that a unary sequence has been received. Thus the unary sequence is independently specified by the prime numbers. Since the unary sequence is both complex and specified, it constitutes CSI. The SETI scientists would have good reason to say that the CSI they received indicates design and would infer that they had discovered traces of intelligent alien life.

The Law of Conservation of Information

Dembski establishes with some rigor that CSI cannot be generated by necessity, chance, or their combination. In outline, his argument is thus.[15] Necessity can be understood as a deterministic mapping, algorithm, or function f that takes some event i as input and computes some other event j as output. That is, $f(i) = j$. Clearly if i fully determines j, then it is not possible for j to contain any information that it did not derive from i. Therefore, in particular, j cannot contain any CSI not contained in i. But in that case f does not explain the appearance of CSI; it is merely a conduit by means of which CSI is moved around. It may be objected that it is not merely the event identified as the cause but also the wider environmental or initial conditions that influence an effect. Might not some novel CSI appear in j in this way? This move does not help. If i does not fully account for j, then it is simply an incomplete specification of the causal resources. Suppose that any other wider factors involved are denoted w, then the new mapping is $f(i,w) = j$; it follows that any CSI in j derives from i or w.[16] The question of where the CSI in i or w originated remains unanswered. Therefore, necessity alone cannot account for the origin of CSI.

Likewise, chance alone cannot account for CSI, such as the CSI in one of Shakespeare's sonnets. Chance can generate simple and specified information, for example, a short line from the sonnet, consisting of a meaningful group of words. Or it can produce complex but unspecified information, such as a randomly generated list of words drawn from the sonnet, or more realistically from a large collection of sonnets, and arranged in random order.[17] But chance cannot account for information that is both complex (highly unlikely) and independently specified. Still, it might be claimed that it is the interaction of chance and necessity that generates CSI, which would constitute the model of natural selection, where random mutations interact with the deterministic laws defining a creature's environment. However, this claim does not help. If neither random events nor deterministic laws generate any CSI, then neither does their combination. If a random event i that contains zero CSI is then algorithmically mapped to another event j, it is still the case that j cannot contain any CSI not contained in i. Since i contains zero CSI, it cannot account for any CSI in j.[18]

These results justify what Dembski calls the law of conservation of information (LCI). Let "CSI (E)" denote the complex specified information contained in an event E, and "Σ_i CSI(NC_i)" denote the sum of the complex specified information contained in all the natural causes NC_i of E. Natural causes here include events that result from chance, necessity, or the interaction of the two. Then,

LCI: If only natural causes NC_i affect E, CSI (E) $\leq \Sigma_i$ CSI(NC_i).

In other words, if only naturalistic causes are operating, then the amount of CSI in an effect cannot exceed the amount of CSI in the sum of the natural causes. While naturalistic causes can transmit, reshuffle, or lose existing CSI, they cannot generate CSI de novo. The general importance of this law is that it sets a limit on legitimate reductionist explanations. As Dembski writes:

> The Law of Conservation of Information . . . shows that CSI cannot be explained reductively. To explain an instance of CSI requires either a direct appeal to an intelligent agent who via a cognitive act originated the CSI in question, or locating an antecedent instance of CSI that contains at least as much CSI as we started with.[19]

While it is not our focus here, LCI provides a principled explanation of the failure of the various Darwinist accounts of irreducible complexity we saw in chapter 4. Irreducibly complex biological structures (at least if they are also "very complex") are ones that contain more CSI than any credible

Darwinian precursor; unfortunately, the Darwinian resources of chance and necessity are unable to make up the difference because they cannot generate any additional CSI.

The kind of skeptic we have in mind doubts that the external world really exemplifies CSI. However, LCI can be applied more specifically to our representations themselves. Let "R $(x_1, x_2, \ldots, x_n)$" denote a representation that represents n objects (real or imaginary) as being a certain way. Then we have

RLCI: If only natural causes NC_i affect $R(x_1, x_2, \ldots, x_n)$, then
$$\text{CSI}(R[x_1, x_2, \ldots, x_n]) \leq \Sigma_i \text{CSI}(NC_i).$$

This principle tells us that if a representation R contains CSI and if the representation is not even in part supernaturally caused, then the amount of CSI in R cannot exceed the sum of the CSI in R's natural causes. This is really just a more rigorous formulation of Dretske's Xerox principle, which argues that representations (or copies) can carry at most the same amount of information as the real-world entities (or originals) that caused them.[20]

While we are at it, we should also consider those cases where a representation causally precede actions, as when builders follow an architect's blueprint or when programmers code an algorithm. In that case, we would seem to require the converse of RLCI. However, whether representations, especially mental representations, can be understood as natural causes is deeply controversial. Yet, it is not controversial that action involves a lot of causal factors that are natural causes, for example, events in the nervous system and body. So it seems best to relate representations to actions as follows. If A is the action, NC_i the uncontroversial natural causes of A, and $\Sigma_j R(x_1, x_2, \ldots, x_n)_j$ is the sum of the j representations that causally contributes to A,[21] we have

$$\text{ALCI: CSI}(A) \leq \Sigma_i \text{CSI}(NC_i) + \text{CSI}(\Sigma_j R[x_1, x_2, \ldots, x_n]_j).$$

The principle ALCI says that an action A can contain no more CSI than that which is contained in the sum of its uncontroversial natural causes and representational causes—which for a naturalist would be just more natural causes.

The principles RLCI and ALCI are all we will need to state two arguments for the legitimacy of design. The first form applies RLCI to our theoretical reasoning, to show that our thoughts themselves evince design. The second form applies ALCI to our practical reason and action, showing that the effects of our thoughts also exemplify design. The connection between thought and action means that design cannot be "domesticated": it is not just a feature of our thought; rather, it spills over into what we call the external world. Information-bearing thoughts leave a trail of information in the world, a trail that can only be read or understood on the assumption that it resulted from an intelligent agent and can only be deciphered by another intelligent agent.

The Argument from Theoretical Reasoning

In chapter 5, we noted that our reasoning exhibits irreducible complexity. The reasons an agent has for a conclusion (or action) are tightly coordinated and well matched. Often, an agent's reasoning also constitutes CSI, which is no surprise if, as Dembski argues, irreducible complexity is a special case of CSI. This is especially clear in complex cases of logico-mathematical argument. For example, consider Kurt Gödel, at the point that he has fully worked out the details of his famous incompleteness theorem. Without doubt, Gödel's proof is highly complex, so Gödel's mind must have contained a highly complex representation. According to our notion of design, this representation constitutes Gödel's design for the proof that he wrote down.[22] Thus, according to our notion of design, these representations really are designs. This assumption is only justified, however, if there are such things as designs. If Gödel's representation arose through chance and necessity, then an eliminativist can claim that it only appears to be a design. To show that we can legitimately think of Gödel's representation as a design, we therefore need to show that it could not have arisen via chance, necessity, or the combination of the two.

So we need to analyze the character of the representation that we claim is Gödel's design for his proof. Certainly, this representation contains far more than 500 bits of information. So the representation is complex. What is more, each step of the argument is justified by a principle of logic, known to Gödel, that is independent of the proof itself. That is, any principle that Gödel makes use of can be used just as well in other proofs. In that sense, Gödel's proof, and hence his mental representation of it, is independently specified by these principles.[23] Therefore, since it is both complex and independently specified, Gödel's mental representation of the proof exhibits CSI.

On the face of it, this mental representation contains "new" CSI, since Gödel's proof is highly original. But given RLCI, new CSI does not have a naturalistic explanation. So, barring inspiration from God or alien mathematicians, it is reasonable to suppose that the agency of Gödel himself explains the new CSI. On a constructivist interpretation, this would be additional CSI, which Gödel generated by constructing new principles of logic. On a Platonist model, this would be additional CSI that Gödel acquired by discovering objective logico-mathematical truths. Either way, Gödel's intelligent agency is implicated in the explanation of the mental representation of the proof. What is more, if Platonism is right and Gödel discovered the principles he used, then this CSI already existed and must ultimately result from some other, higher, intelligent agent. It is not surprising that some mathematicians, Gödel included, thought of themselves as trying to read God's mind. However, even if Gödel did construct the principles, the ability to do so required cognitive abilities that themselves exhibit CSI. Although they may have been improved by Gödel's own industry, these cognitive abilities were not created by Gödel's agency, and so the CSI they contain traces to another source. This source would then eventually be implicated if we try to explain the principles Gödel constructs. Thus, even a constructivist Gödel could give glory to God for his work.

However, although very unlikely, it might be that the CSI in Gödel's mind all derived from preexisting innate principles of his, and hence the representation contained no "new" CSI—that is, no CSI new to Gödel's mind. Yet, this only displaces the naturalist's problem. His new problem is to explain the origin of the CSI in the innate principles Gödel used. But, given RLCI, this too cannot have a naturalistic explanation. Natural causes can only explain how CSI can be shuffled about; only intelligent agency can account for the appearance of CSI in the first place. So again, even if Gödel's principles are innate, some other intelligent agency is implicated.

It follows that regardless of whether or not Gödel's mental representation of his proof contains CSI that is new to Gödel, its existence implies that design is a real category. This shows that our notion of Gödel's mental representation as a design for his proof is legitimate. So, at least in cases where a putative design exhibits CSI, design is a legitimate category.

The Argument from Action Explanation

When we think of an action, we think of it as a sequence of movements that implements some prior design. Is this way of thinking of action legitimate? It is only if we can show that actions really do conform to designs and that design is necessary to explain some objective characteristic of our actions. However, at least for complex actions, this is certainly the case.

Suppose an agent implements a complex plan. Then his intentional action exhibits CSI. For example, if a scientist carefully plans a complex experiment and then follows the plan, the scientist's action is necessarily both complex and specified. The complexity is inherited from the complexity of the plan for the experiment and no doubt supplemented by unforeseen complexities as well; the plan itself is the independent specification of the action.

The principle ALCI helps us to see why physicalist action explanations, either eliminativist or naturalizing, fail to account for the scientist's performance. According to the physicalist, what the scientist does must be explained in terms of the natural processes going on in the brain, nervous system, and the rest of the body. However, when these are given physical descriptions, they omit the CSI that is contained in the mental representations of the scientist's plan. Unsurprisingly then, when the physicalist attempts to explain the scientist's action, all the scientist can account for is a variety of physically described movements. These movements can have only as much CSI as in their physically described causes. Since the contribution of the CSI of the mental representations was ignored, the CSI of the intentional action description cannot be accounted for.

What ALCI tells us is that you need to include the CSI in the representations that cause an action if you want to fully account for the CSI exhibited by the action itself. If you throw out the information in an action plan, you are unable to explain the action as an implementation of that plan. Actions convey information. This holds not only for the obvious cases of speech and writing but also for any intentional action that implements a plan. If we have some grasp of that plan, if we know something about what the agents are trying to do, we can acquire more information about their intentions by watching what they do, which explains why one way to learn a complex task is to observe someone else performing it. For example, we learn that a carpenter is going to make a chair. By careful observation, we learn how the carpenter has planned the preparation and assembly of the pieces. The information conveyed by the action does not arise magically. It is not like rain produced by a rain dance. On the contrary, it is clear that actions inherit at least some of their CSI from the mental representations that cause them, which is why when we are unclear about what agents are doing, we are well advised to ask them what they are trying to do. To do so gives us the needed specification of the action so that we can decode the CSI that was all the while present. Because actions inherit their CSI from the representations that cause them, actions can never be fully explained by appeal to physically described events in the brain and body of an agent. In this way, applying CSI to action explanations not only vindicates the basic stance of folk psychology and thoroughly refutes behaviorist and eliminative approaches, but it also helps to explain why folk psychology works. The reason that appeal to agents' representations helps to explain their actions is that those representations decode critical information that the action contains. Folk psychology is thus not "rationalization" in some discredited Marxist or Freudian sense but the art of decoding or interpreting (frequently) nonlinguistic information. Action explanation is a form of hermeneutics.

It follows that our notion of complex actions depends on the notion of CSI. None of the CSI in an agent's complex action can be generated de novo from the natural processes going on inside the agent's brain and body. Either this CSI derives from an agent's own intelligent agency, or it was acquired from elsewhere. For example, students are given directions when performing well-tried chemistry experiments. The CSI in the students' actions derives from mental representations that acquired their CSI from the directions. The directions derive in turn from the instructor's mind, which may have originated the directions, found them in a book, or been taught them by someone else. But an infinite regress is here impossible. Eventually, we must find some origin of the CSI; given LCI, this cannot be chance, necessity, or the interaction of the two. Only intelligent agency can explain the appearance of CSI de novo.

So again, when we think of actions as designed, this way of thinking is demonstrably legitimate, so long as the actions are sufficiently complex. But that means that design is not just an attribute of our thoughts. It spills over into the effects of those thoughts in the world around us. But if that is so and we can detect design in human actions, we should be open to the possibility that we can detect design in nonhuman actions as well.

Objections and Replies

One can imagine Dennett's objecting that CSI might simply derive from Mother Nature. However, as we saw, "Mother Nature" is an ambiguous term. If Mother Nature simply means natural selection, then it is at best a conduit for existing CSI. It cannot account for the origin of any new CSI. This is an example of what Dembski calls the "displacement problem."[24] Claiming that CSI derives from a source unable to create it merely displaces the problem of where it originated. If CSI is indeed shuffled around by natural selection, then it remains unexplained why CSI was available in the first place. Yet, if Mother Nature is described as an intelligent agent, it is at least understandable how she could originate CSI. However, as we saw in chapter 3, viewing Mother Nature in this way is flatly incompatible with scientific materialism. Either way, there is no plausible explanation of the origin of CSI.

Another approach would be to claim that CSI is inherited from the laws of nature. This suggestion is made via Dretske's account of the origin of intentionality in information:

> The ultimate source of the intentionality inherent in the transmission and receipt of information is, of course, the nomic regularities on which their transmission of information depends.[25]

It is clear, however, that locating intentionality in the laws of nature only displaces the problem of where intentionality originated. Similarly, the fact that laws of nature themselves exhibit CSI is not an ultimate explanation for the existence of CSI. On the contrary, since natural laws pervade the universe and exhibit CSI, it is clear that CSI cannot have a naturalistic explanation. Either CSI has no explanation at all—that is, it is a brute fact—or the cause of CSI must be something independent of a law-governed universe. It is, however, highly implausible to claim that CSI is a brute fact since, by definition, events that exhibit CSI are of very low probability and conform to an independently specified pattern. Saying that the CSI in natural laws is a "brute fact" can only mean that it arose by chance,[26] but CSI eliminates the chance hypothesis, so this cannot be maintained. Since the only known origin of CSI in our experience is intelligent agency and since no natural cause can account for the CSI that pervades the universe, we should conclude that some supernatural agency is the ultimate explanation of CSI.

Conclusion

In this chapter, I have fleshed out two main arguments for the legitimacy of design. The first shows that the very idea of a concept is linked to intentionality and design. If there are such things as concepts, then certainly the concepts of intentionality and design are not fictional. The second argument shows that even if concepts are jettisoned in favor of an information–theoretic approach to the mind (not a coherent choice in my view), the CSI contained in plans for logico-mathematical proofs and in complex actions can only be accounted for by literal design. Granted this Kantian deduction of design, physicalist attempts to eliminate or naturalize design and intentional categories must fail. If intentionality and CSI are indeed sui generis, irreducible to the natural categories of chance and necessity, then the only ultimate explanation is supernatural agency. Either scientific materialism fails to explain intentionality and CSI, or it explains them by abandoning materialism.

Notes

1. Dretske, *Knowledge and the Flow of Information*, 76.
2. Dembski, *No Free Lunch*, 206–7.
3. Thus, it is no use explaining someone's staying indoors by attributing the belief that it is raining, if in fact the individual desires to get wet but has to finish a tax return.

4. Crane and Mellor, "There Is No Question of Physicalism," 200.
5. Paul Draper, "Irreducible Complexity and Darwinian Gradualism: A Reply to Michael J. Behe," *Faith and Philosophy* 19, no. 1 (January 2002): 3–21.
6. Dretske, *Knowledge and the Flow of Information*, 51.
7. That is, there are 100 alternative scores, in the ranges 0–91 and 93–100, if we assume no fractional points and no extra credit.
8. Dretske, *Knowledge and the Flow of Information*, 8.
9. Dembski, *No Free Lunch*, 126.
10. "For a thoroughly shuffled deck of cards, the probability of being dealt a royal flush . . . is approximately .000002 whereas the probability of being dealt anything other than a royal flush . . . is approximately .999998" (Dembski, *No Free Lunch*, 126).
11. For further discussion see Dembski's *No Free Lunch*, 127ff.
12. See Dembski, *No Free Lunch*, 128.
13. Dembski, *No Free Lunch*, 15.
14. Dembski, *No Free Lunch*, 142–45.
15. For full details, see *No Free Lunch*, 3.8, 149–59.
16. As Dembski points out, one can even account for any information content in the mapping relation f itself, by using a universal composition function U, and setting $U(i,f) = f(i) = j$. See Dembski, *No Free Lunch*, 154–55.
17. Strictly speaking, computer random-number generators are only pseudorandom, using an algorithm to simulate the appearance of truly random sequences in a finite case. To have a truly random case, game show contestants would have to be selected on the basis of the amount of radiation given out by a decaying nucleus or some other truly random event.
18. As before, if there is any CSI in j, it derives from the function f itself or from some nonchance input that already had some CSI of its own. However, this does not solve but merely displaces the problem of how CSI was generated in the first place.
19. Dembski, *No Free Lunch*, 164. Dembski's technical formulation of the law of conservation of information is different, but the basic idea is the same.
20. Dretske, *Knowledge and the Flow of Information*, 57.
21. This does exclude epiphenomenalism, but we have already argued that this doctrine implies that it is very unlikely that our beliefs are usually true and our desires are usually appropriate (chapter 6).
22. Note that it is irrelevant whether Gödel could be conscious of the entire representation of the proof at any one time.
23. Granted a Platonic understanding of logic and mathematics, there is an even stronger sense in which the principles Gödel used are detachable: they have mind-independent, objective reality and are discovered, not constructed, by the mathematician.
24. Dembski, *No Free Lunch*, 4.7, 203–7.
25. Dretske, *Knowledge and the Flow of Information*, 76.
26. It won't help to say that it arose from necessity, since that would merely add more laws exhibiting CSI and since it is this characteristic of laws that requires explanation.

CHAPTER EIGHT

~

Science and Christianity: Dogmatism and Dialogue

> He had in fact made up his mind beforehand, and did not properly consult experience as the basis of his decisions and axioms; after making his decisions arbitrarily, he parades experience around, distorted to suit his opinions, a captive.[1]

> Unless we are willing to suspend . . . explanations at the particular points where these explanations are inappropriate to the particular data, we in principle eliminate even the possibility of discovering anything new. . . . The particular must triumph over the general, even when the general has given us immense help in understanding the particular.[2]

It is customary to think of dogmatism as a problem of religion and to suppose that science is essentially an enlightened mode of free enquiry. In fact, the process of secularization has fueled a religion of scientism no less dogmatic than the worst excesses of organized religion. In the following section, I argue that Darwinism in particular, at least as it is understood by scientific materialism,[3] has become a new kind of scholasticism, differing in content but protected from serious criticism in much the same way as its medieval precursor. Indeed, many of Bacon's trenchant criticisms of the Aristotelian paradigm for science apply with equal force to the deductionist excesses of Darwinism. But the analogy has a second and more disturbing parallel. The great synthesis of Aquinas between Christianity and Aristotelian science and philosophy tended to invest the latter with the religious significance and authority of the former so that anyone who dissented from Aristotle was in danger of being branded a heretic. This confusion of science with religion continues, except that now the religion is not Christianity but the naturalistic religion of scientism. It is therefore unsurprising that some of the Darwinist faithful perpetuate a climate of ideological intolerance, replete with orthodoxy tests and even inquisitions (to be discussed later in the chapter).

It is also unsurprising that many Darwinists—such as Richard Dawkins, Daniel Dennett, and William Provine—think that Darwinism precludes any hope for the peaceful coexistence between science and religion. If Darwinism is inextricably wedded to a metaphysical naturalism that logically excludes supernaturalistic religions such as Christianity, and if, at least at present, being scientific means being a Darwinist, then, again, no scientist should be a Christian. Not everyone agrees with this stark conclusion. Michael Ruse, for one, has argued that Darwinists can be Christians so long as they are strictly methodological naturalists, pursuing science as if nature is the final frontier but allowing the possibility that it is not. Ruse has also shown great courage and determination in promoting meaningful dialogue between Darwinists and their critics, sometimes being stoned by all sides for his trouble.[4] For this chapter, I evaluate Ruse's attempt to show the compatibility of Christianity and Darwinism. I argue that, although well intentioned, it fails to fully achieve its aim because of its (perhaps unnecessary) commitment to reductionism. In the final section of the chapter, I present my own view of what it would be like for Christianity and science to participate in meaningful dialogue.

The New Scholasticism

"Scholasticism" has both a positive and a negative connotation. At its best, scholasticism produces rigorous philosophical demonstrations, such as Aquinas's Five Ways to God. At its worst, scholasticism means a constant polishing and recycling of old ideas and secondary sources (such as commentaries), which inhibit both the discovery of important new truths and the recovery of original texts. I will use the term "scholasticism" more narrowly, to signify a particular dimension of the negative connotation—that is, the flawed attempt to extend knowledge by uncritically affirming the logical consequences of preconceived opinions. This sort of scholasticism is precisely what Francis Bacon was criticizing in *The New Organon*.[5] According to Bacon, both Aristotle and his followers thought that once one had intuited the essence of a kind, it was merely a matter of deduction to discern how particular examples of the kind must behave. This view may be called *uncritical deductionism*, in contrast to Bacon's more cautious "true induction" or Karl Popper's method of conjectures and refutations, either of which might aptly be called *critical deductionism*. To be sure, Bacon paints with a broad and polemical brush. The scholastic period did have respectable and competent scientists who were not uncritical deductionists, for example, the great mathematical physicist Buridan. Likewise, when we turn to the parallel with Darwinism, it would be wrong to claim that all Darwinists are like the worst scholastic thinkers: the best are both self-critical and tentative in their claims to knowledge. What I argue is that uncritical deductionism has become a feature of dogmatic Darwinism, a stance that makes claims to certainty in areas where more cautious scientists would be content to claim they had working hypotheses. In the worst cases, strong factual objections that ought to count against the official Darwinian explanation are simply overruled by the prior conviction that the theory must be right despite the appearance to the contrary.

To be sure, it is well known that a degree of tenacity in the face of objections can be vindicated in science. When Newtonian theory was unable to explain the perturbations in the orbit of Uranus, it would have been rash to scrap a theory that had been so successful in other areas. And, as it turned out, the discovery of Neptune and its gravitational interaction with Uranus sufficed to explain away the anomaly. But in this case, Newton's theory made a specific prediction concerning the mass and whereabouts of a planet, which, if it existed, would account for the anomaly. This is frequently not the case with the claims of dogmatic Darwinists. We are not told where to look for specific factors that could account for the discrepancy of observation with theory. Instead, we are simply told that some or other such factors must exist because we already "know" that the theory is true. It is this lack of specificity in the predictions of Darwinists that makes it easy for them to fall into the scholastic trap of uncritical deductionism.[6]

The best recent illustrations of my thesis are found in Jonathan Wells's *Icons of Evolution*.[7] In this powerfully written book, Wells critiques ten of the leading "icons" that have been used in school and college textbooks to support Darwinism. An icon is a representation or likeness, so it is something either true or false, helpful or unhelpful. Icons are not necessarily bad, but they can exaggerate the certainty of science because such pictures create the impression of established fact where in reality there may be only plausible conjecture. An icon can also be a devotional picture, an expression of religious or ideological veneration. In that case, misplaced zeal may enshrine an icon that falsifies reality. The icon may become what Francis Bacon called an "idol" of the mind, a prejudice that obstructs understanding of the natural world. Most dangerous are what Bacon calls the "idols of the theatre," cognitive illusions derived from "the various dogmas of different philosophies and even from mistaken rules of demonstration" leading to "so many [stage] plays produced and performed which have created false and fictitious worlds."[8] Wells's thesis is precisely that the icons of evolution have become idols that distort the facts and lead to "fictitious" explanations disconnected from reality.

Indeed, this is where the connection to scholasticism becomes clear because Wells's critique of dogmatic Darwinism parallels Bacon's attack on scholastic science. Bacon advised that "nature is conquered only by obedience."[9] Obeying nature means listening to what it is really saying. Scientists should not be tied to a dogmatic starting point, such as Dobzhansky's maxim that "nothing in biology makes sense except in the light of evolution."[10] Rather, they should assert that "nothing in biology makes sense except in the light of evidence."[11] Bacon blasted the Aristotelian approach because it "flies from sense and particulars to the most general axioms, and from these principles and their settled truth, determines and discovers intermediate axioms."[12] This is precisely the criticism that Wells levels against the Darwinist who "starts with a preconceived idea and distorts the evidence to fit it."[13] As Popper and Lakatos so strenuously argue, real science does not immunize a theory from falsification. Rather, it shatters myths that attempt to anticipate nature, and it lets the evidence speak. This tension between good and bad science, between critical and uncritical deductionism, is illustrated by each of the ten icons Wells examines. Since Wells has already presented his case at length, I refer the reader to his book for the details. For this section, I simply point out the ways in which they give evidence of the scholastic tendencies of dogmatic Darwinism.

In 1953, the Miller–Urey experiment seemed to show how the building blocks of life could arise from lightning in Earth's primordial atmosphere. Diagrams of the apparatus used are a continuing icon in contemporary college textbooks, yet geochemists have known for over ten years that the experiment makes false assumptions about Earth's atmosphere. The experiment only works in the absence of oxygen, but geochemists have found strong evidence in the composition of rocks that oxygen was present.[14] Instead of accepting this difficulty, dogmatists have claimed that since chemical evolution of life must have happened this way, the early atmosphere must have lacked oxygen! As Bacon put it in his day, "Current logic is good for establishing and fixing errors . . . rather than for inquiring into truth."[15]

Textbooks also portray Darwin's tree of life, showing all species as the result of descent with modification from common ancestors. Unfortunately, this icon has been dislodged by fossil and molecular evidence. The fossil record shows an extraordinary explosion of distinct body plans in the early Cambrian, which has not changed with the discovery of pre-Cambrian fossils. Attempts to save Darwin's tree using molecular evidence have shown instead that life is better represented as a complex thicket.[16] If Darwinism were really advancing a testable theory, these facts would falsify its current formulation and require significant modifications at the very least. Instead, they are treated as "problems" that will eventually be explained away. This kind of answer is uncomfortably reminiscent of Marxism: when predictions of capitalist downfall repeatedly failed, an endless cast of extenuating circumstances was paraded before us.

Another favorite image depicts homologous limbs in vertebrates. Empirically defined, "homology" means structural similarity, and it is a fact that the bones in the forelimbs of bats, porpoises, horses, and humans are remarkably similar. Unfortunately, rather than considering a range of possibilities, dogmatic Darwinists insisted that the only possible explanation for homology was common descent. Very soon homology was defined as similarity due to common descent. This scenario resembles a bad detective whose first question is how a dead person was murdered, as if no other explanations for death are possible. If we define something in terms of its presumed etiology, we exclude alternative accounts of origins by fiat. Yet Darwinists continue to claim that homology is evidence of common descent, that is, that a result of common descent is the result of common descent.[17] Such a tautology not only begs the question against rival explanations, but it also cannot be tested, because even in principle, cases of homology could not exist that did not result from common descent. This tendency to the untestable supports the view once expressed by Karl Popper, that Darwinism is really a metaphysical research program, rather than a scientific theory.[18] Popper's point was not that Darwinism was worthless but that a metaphysical research program that implies only that some or other explanation must be true should not be confused with a specific, testable scientific theory. Dogmatic Darwinists are clearly guilty of ignoring this distinction, appealing to prior metaphysical prejudice to establish "scientific findings," even when they lack a basis in independent, empirical fact.

An idea may be powerfully evoked even if it is not explicitly portrayed. Haeckel, a contemporary of Darwin, made famous drawings of embryonic development. The drawings represent ontogeny, the development of individuals; but they also suggest the biogenetic law, that ontogeny recapitulates phylogeny, that is, the evolutionary history of the species. According to Haeckel's drawings, the embryos of fish, salamanders, tortoises, chicks, hogs, calves, rabbits, and humans have virtually no differences when in the early stages of development. This is what Darwinism would predict, since it claims these species arose from common descent. However, we now know that the early embryos are very different[19] and that Haeckel's drawings were faked. Scientists of the time were not eager to check out Haeckel's work, so the biogenetic law was "deduced from evolutionary theory rather than inferred from the evidence."[20] And even today, "biology textbooks continue to teach it."[21]

Darwin's theory of gradualistic evolution predicts a multitude of transitional forms; and when they are not found, Darwinists write it off as an imperfect geological record.[22] Still, some are needed, and *Archaeopteryx*, discovered in 1861, had seemed to be the clincher. In fact, modern paleontologists do not think this bird was the ancestor of any modern bird, and its own origins are disputed. Darwinists have tried to find better fossils, and their desperation has led to credulity. In 1999, scientists announced the discovery of *Archaeoraptor*, which turned out to be "a dinosaur tail glued to the body of a primitive bird."[23] Later, a sample of triceratops DNA was trumpeted to be similar to the DNA of modern turkeys. Actually, it was 100 percent identical, prompting the suggestion that the sample had been contaminated by a turkey sandwich. "This isn't science. This isn't even myth. This is comic relief."[24] Of course, even competent scientists make mistakes. But the almost completely uncritical acceptance with which some scientists greeted these findings is strong evidence that their minds were already made up and not to be confused with the facts.[25]

Still, until recently, it had been thought that two icons were unassailable: the peppered moths and Darwin's finches. But even these face strong objections.[26] The claim was that darker peppered moths would have an adaptive advantage if they rested on tree trunks blackened by pollution since the moths would be better camouflaged from predators. The problem is that peppered moths don't rest on tree trunks but hide under branches. Like the dead parrot of Monty Python that is nailed to its perch, textbook pictures show dead moths stuck to tree trunks. That Wells is correct in this assessment has recently been powerfully corroborated by a meticulous book length study by Judith Hooper.[27] By contrast, Darwin's finches do provide evidence of natural selection. During a drought on the Galapagos Islands, the average beak size of finches increased, enabling them to eat the larger, drought-resistant seeds. By extrapolation, it was claimed that a new species of finch might develop within two hundred years. Unfortunately, when the rains came, the beak sizes returned to normal, and the evidence supports only oscillating natural selection with no net evolutionary change.

It is well known that bacteria gain resistance to antibiotics and that insects adapt to insecticide. However, these changes do not affect morphology: they do not create a new body plan, which is what is needed for a new species to arise. It seemed that fruit flies would change all this. With a series of controlled mutations, it is possible to develop fruit flies with four wings rather than two. Unfortunately, the new wings lack flight muscles and replace the fruit fly's stabilizers, so the result is less well adapted.[28] The claim that such mutations provide evidence of large scale "macroevolution" depends on presupposing relevant beneficial mutations, even though these are not what empirical investigation has turned up. Incredibly, the assumption that such mutations must exist is treated as equivalent to our having evidence for their existence.

At one time the fossil record seemed to show a clear, linear transition from ancient four-toed horses to modern single-toed horses.[29] Later, it seemed there was more branching,[30] and this was used by neo-Darwinists to claim that evolution is an undirected process. This position tells us that Darwinism is consistent with both strong linearity and strong nonlinearity. It is therefore obviously going to be difficult to find anything that could count as evidence that Darwinian evolution had not occurred. Further, the claim that evolution is undirected is a metaphysical thesis, not part of empirical science. Indeed, since Darwinists have already assumed that evolution is undirected, it is hardly something they can claim to learn from their observations.

Even more dogmatic, and an apt example of Bacon's idols of the theatre, are the accounts of human evolution offered by paleoanthropologists. Here it becomes obvious that certain stories are appealing for ideological reasons and that, because fossils are not self-interpreting,[31] they must be artistically placed "into preexisting narrative structures."[32] As Constance Holden wrote in *Science:*

> The primary scientific evidence is a pitifully small array of bones from which to construct man's evolutionary history. One anthropologist has compared the task to that of reconstructing the plot of *War and Peace* with 13 randomly selected pages.[33]

Given the vast number of alternative theories consistent with the same facts, it is entirely possible that any such narrative is as divorced from reality as an imaginative stage play. Yet, because of the certainty that the background evolutionary theory is true, the intuitive plausibility of a favored narrative makes some observers treat it as scientific knowledge.[34] One can hear Winston Churchill as a biologist: "never in the field of science have so many based so much on so little."[35]

Wells's book has elicited a plethora of caustic reviews, but as Wells himself has shown in detail,[36] very few of them affect the substance of his claims. It seems to me, therefore, that Wells's book amply documents the charge that dogmatic Darwinists are prone to uncritical deductionism and that scientists who are serious about objectivity and testability should insist on a higher standard.

Everyone Expects the Darwinian Inquisition

Even the most dogmatic Darwinists never engage in the terrible acts of cruelty associated with the medieval inquisitions. Nonetheless, in a more civilized and urbane way, some Darwinists within the scientific establishment have attempted to intimidate and censor their critics. In its milder forms, this amounts to little more than the sort of dismissive, ad hominem abuse that, lamentably, has spread from the political realm into the academy. For example, Richard Dawkins found the intellectual generosity to opine:

> It is absolutely safe to say that if you meet somebody who claims not to believe in evolution, that person is ignorant, stupid or insane (or wicked, but I'd rather not consider that).[37]

Likewise, Dennett shows his concern for the religious critics of Darwinism by offering them the hope that their places of worship will be preserved as "cultural zoos."[38] In a society obsessively worried about hate speech, it is heartening to learn that there is no such thing as bigotry against anti-Darwinist infidels. Only this prejudice can explain some of the cruel, inaccurate, and logically irrelevant accusations that were leveled against Jonathan Wells after publication of *Icons of Evolution*. Wells has documented the ways in which his critics have portrayed him as ignorant, stupid, *and* wicked.[39]

But it has not stopped at abuse. Reputations are attacked; positions are denied; and serious scientific work that challenges Darwinian orthodoxy is excluded from science journals.[40] One could write a whole book on the topic. But for now, I simply consider some salient case studies.

In 1990, when the accomplished science writer and inventor Forrest Mims admitted that he questioned evolutionary theory, he was not hired to write the Amateur Science column for *Scientific American*. In the ensuing protest, which included many voices opposed to design but even more opposed to viewpoint discrimination, the journal *Science* printed the following:

> Even today, some members of the scientific establishment have seemed nearly as illiberal toward religion as the church once was to science. In 1990, for instance, *Scientific American* declined to hire a columnist, Forrest Mims, after learning that he had religious doubts about evolution.[41]

Similarly, in 1992, Dean Kenyon, a biology professor at San Francisco State University, was barred from teaching introductory biology classes after he shared his misgivings about evolutionary theory, including his own theory of chemical evolution.[42]

> Mr. Kenyon . . . had for many years made a practice of exposing students to both evolutionary theory and evidence uncongenial to it. He also discussed the philosophical controversies raised by the issue and his own view that living systems display evidence of intelligent design. . . . He was yanked from teaching introductory biology and reassigned to labs. . . . Fortunately, San Francisco State University's Academic Freedom Committee . . . determined that . . . a clear breach of academic freedom had occurred.[43]

In 2000, William Dembski, one of the leaders of the intelligent design movement, was removed from his position as director of the (now defunct) Michael Polanyi Center, at Baylor University. As I have argued elsewhere,[44] there is no question that this decision was made to placate hostile faculty, concerned for their academic reputation among secular institutions. Apparently, even at a Christian Baptist school like Baylor, conformity to secular canons of academic correctness is more important than a genuinely open discussion of the scientific case against Darwinism and in favor of intelligent design.

Can a Darwinian Be a Christian?

Dismayed by the hostility that arises when strong Christians confront the claims of materialistic science, it is natural to hope for some sort of accommodation.[45] Disliking friction, many favor a complementarity thesis. In the formulation of the late, noted paleontologist Stephen Jay Gould, science and religion define nonoverlapping *magisteria*, with science the authority over observation and fact, and religion the magistrate on matters of ethics and ultimate meaning. It is covertly assumed that only science deals with objective truth, suggesting that the "values" of religion are only private preferences. In practice, complementarity approaches demand what Sir John Templeton has called "humility theology"[46] but without "humility science." However delicately the matter is put, imperialist "science"[47] is expanding its empire and telling religion what it can have as leftovers.[48]

Nevertheless, even among the Darwinists are those who hold a more sophisticated view that is less condescending in its treatment of the claims of Christianity. In his *Can a Darwinian Be a Christian?*[49] Ruse states a balanced and ambitious project: "Can someone who accepts Darwin's theory of natural selection subscribe at the same time to the essential claims of Christianity?"[50] Encouragingly, he also says that he will try to consider each pole of the dialectic in robust varieties: "My inclination . . . will be to compare a fairly strong form of Darwinism against fairly traditional forms of Christianity."[51] Ruse realizes that it is too easy to answer the title question in the affirmative if we have a liberal Christianity or a watered-down Darwinism. Ruse is faithful to his announced aim and struggles through the apparent conflicts, looking for unsuspected agreement. In the process are some fine discussions of difficult and important matters.[52] Yet, as we shall soon see, Ruse thinks that if conflict is unavoidable, reductionism trumps traditional Christianity and only liberal theology is on offer.

Some of the standard evidence Ruse cites for Darwinism is dubious,[53] but that is irrelevant since the issue is whether someone who accepts the case for Darwinism can have a coherent Christian worldview. This question is worthwhile, and Kenneth Miller, a noted biologist at Brown University, has made a heroic attempt to answer in the affirmative,[54] arguing that evolution is not only compatible with Christianity but also at times supportive of its claims. Unfortunately, I think that what Ruse's book really does is answer a different though related question. Furthermore, I think his affirmative answer to that question is mistaken. The slippage occurs because Ruse takes Darwinism to be a package deal, including some rather strong background philosophical assumptions. Ruse realizes the obvious incompatibility of strict naturalism with orthodox Christianity, but he claims that science is committed at least to methodological naturalism. As controversial as even this position is, Ruse goes further and asserts that Darwinism is committed not only to methodological but also to ontological reductionism, applicable all the way up to human thought.[55]

Thus, by the time Ruse has finished unpacking his understanding of Darwinism, the question he really addresses is this: "Can a methodological naturalist and full-blown reductionist who believes that natural selection fully accounts not only for the origin of species but also for human nature and cognition (including moral and religious beliefs) have a coherent Christian worldview?" One might ask whether such a strong reductionism leaves room to be merely a methodological naturalist: if everything significant can be reduced to natural categories, won't the supernatural be a redundant rider? Ruse seems to have substituted the question "Can E. O. Wilson be a Christian?" for the original. Amazingly, Ruse's answer to the new question is "Absolutely!" although it won't always be easy.[56] Those of us who think full-blown reductionism is both incompatible with Christianity and bad for science naturally suspect that something got lost in the wash. In what follows I argue that this suspicion is well founded. The problem areas on which I focus occur in Ruse's treatment of miracles, rationality, and sociobiology.

Miracles and Laws

Ruse considers a number of approaches to miracles. He argues that some miracles can be viewed naturalistically[57] while others, if taken literally, would violate the laws of nature. Ruse takes the stance typical of deism that although God could break the laws of nature, it would be aesthetically distasteful for Him to do so, suggesting an incompetent craftsman who must tinker with his handiwork. Ruse then points to the "devastating attacks on law-breaking miracles"[58] of the eighteenth century.[59] His position is that these "law-breaking miracles" are flatly incompatible with science and that traditional Christianity must simply give way. He even suggests a naturalistic account of the resurrection:

> One can think Jesus in a trance, or more likely that he really was physically dead but that on and from the third day a group of people, hitherto downcast, were filled with great joy and hope. . . . It is from this regeneration of spirit that true Christianity stems, not from some law-defying physiological reversals.[60]

On this matter, Ruse could not be more wrong. For one thing, both the incarnation and resurrection are miracles that define the very essence of Christianity, and neither of these miracles can be understood without appeal to the supernatural. The incarnation is God, a supernatural being becoming man. As C. S. Lewis argued, Christianity is one of the few religions that cannot dispense with the concept of a supernatural miracle.[61] If Ruse thinks otherwise, then all he can hope to show is that a Darwinian can be a deist, which will surprise no one. Further, the resurrection is the supernatural emergence of a new glorified form of bodily existence.[62] This miracle was the Father's attestation that Jesus had lived the perfect life and paid the penalty for all our sins. If it did not happen, then we have no basis for our salvation since Christ must have failed in his atoning work. As Paul says, "If Christ has not been raised, your faith is futile; you are still in your sins" (1 Cor. 15:17).

But is Ruse right that such miracles violate laws of nature? Not on any of the standard interpretations of natural laws, as C. S. Lewis and William Lane Craig have shown.[63] On regularity theories, laws are just observed regularities; any alleged "miracle," however unexpected, would at most show that what we thought was a regularity was not. Even if the event could not be accommodated in any obvious way, it would not matter, since such laws have no modal force: they say what has happened and what we expect, not what must happen. Again, one might hold a natural propensity theory that reduces laws to the dispositions or causal powers of certain natural kinds. It is entirely consistent with this picture that while the natural disposition of well water is to stay water, a supernatural addition to that disposition might result in wine.

Ruse, however, clearly thinks that laws are necessary connections. According to this view, laws support counterfactual claims about what would happen under different conditions.[64] Even then, it is a mistake to think that supernatural intervention must break the laws. As Lewis pointed out, it is a superstition to think that laws make things happen. Laws are merely conditionals. A law L1 may say that if an *A*-type event occurs, then a *B*-type event follows. Should a supernatural addition to an *A* make it an *A**, it would not violate L1 if a *B**-event occurred instead of a *B*-event. Indeed, it might be another law, L2, that if an *A**-event occurs, then a *B**-event follows; so that after the miraculous change to A, nature would "resume normal service" with predictable outcomes. Don't we all think that the miraculous wine at Cana made the wedding guests jolly?

More than that, even if an *A*-event does occur and nothing is superadded to it, a *B*-event might fail to occur without violating the law because laws of nature tell us only what will happen when given certain initial conditions. I am playing pool with Ruse. Newton's laws tell us that if his cue ball hits the red at a certain angle and velocity, it will end up in the corner pocket. But when I say, "Michael, your fish and chips are here!" he looks around, and I pick up the red before it is sunk. The effect did not happen, but none of Newton's laws were violated. Instead, the initial conditions were changed. What made this possible is that the billiard table is not a closed system: an outside agency can influence what happens in it.

The orthodox Christian thinks that the whole universe is an open system. Consequently, even if laws are necessary connections, God can intervene to alter either the cause or the initial conditions without breaking those laws; and indeed, the further results of His intervention may themselves be lawlike.[65] Thus, it is not science or its laws that are incompatible with supernatural intervention. The incompatibility is between Ruse's reductionism and miracle. Reductionism requires that "miracles" be definable in terms of law and natural categories alone. This cannot be achieved if the antecedents of laws or their initial conditions can be supernaturally altered. In other words, Ruse's reductionism presupposes that the universe is a closed system,[66] leaving deism as the only theistic option.[67] But deism is not Christianity.[68]

Here it is telling that even a Darwinian such as Kenneth Miller, who is not a reductionist, can believe in the miracles defining Christianity; but a reductionist such as Ruse cannot, because he cannot allow that nature is unable to account for all that goes on within it.

Reason and Nature

One of the strongest objections to reductionist forms of Darwinism is that they cannot explain the character and reliability of thought, an objection I have already developed at considerable length, especially in chapters 5, 6, and

7. C. S. Lewis raises the general worry that if science reduces reason to nonrational processes, it will undermine its own rationality.[69] In a similar vein, Alvin Plantinga has argued that evolutionary naturalism is self-defeating in that, if true, it would remove any ground for relying on human reasoning, including the reasoning used to defend evolutionary naturalism.[70] Thus, were evolutionary naturalism true, we could never have valid reason to believe it because the theory is its own defeater.

Plantinga's point is that natural selection selects for features that enable us to survive but that we could have such features without having reliable thought processes. Might not pleiotropy (one gene coding for multiple features) mean that a gene provides an automatic response to stimuli with survival value and produces conscious thought as an accidental by-product with no such value? Then thoughts would be irrelevant epiphenomena, as if my body were responding to a crocodile infested swamp while I think I am "boozing it up with Freddie Ayer" at an Oxford dinner![71]

Ruse jumps on Plantinga's seemingly silly example and argues that of course accurate thoughts would help me negotiate the swamp and hence could be selected for. But I think Ruse has missed an important point originally made by C. S. Lewis.[72] Lewis points out that what selection preserves is advantageous responses to the stimuli of a given environment and that it works in a thoroughly mechanistic fashion. But this process can extend indefinitely without requiring anything like human thought: "The relation between response and stimulus is utterly different from that between knowledge and the truth known."[73] This is so because effective responses can be developed without developing such a thing as the conscious belief that the response was effective.

As we saw in chapter 6, this can be illustrated by viewing an advanced self-correcting game-playing program as analogous to precognitive "Skinnerian creatures." To be generous to such creatures' powers, suppose they are analogous to an artificially intelligent chess program. Then we can regard each bout of chess as the lifetime of one organism, where "survival" and "reproduction" (a new bout of play with an adjusted algorithm) depend on winning a certain percentage of the games.[74] The process is entirely mechanistic. In each of its generations, the responses of the program become subsequently better attuned to its environment until it conquers even grand masters. But unlike the masters, the program never has thoughts, far less accurate ones, about what it is doing. Nor does it reason to a conclusion or see that a certain outcome is probable or necessary if it follows certain "actions." And it does not really act either, having no goals of its own.[75]

As I have already argued at length, the reason such selection never produces thought is that the intentionality of thought cannot be generated by blind mechanistic processes.[76] None of the latter events are about anything, nor do they have the direction and goal of logical thinking. Furthermore, they are all contingent or, at best, nomically necessary events. But human thought allows us to see that certain propositions are necessarily true ($2 + 3 = 5$) and that a conclusion necessarily follows.

Since nothing in unthinking mechanistic nature is necessary or goal oriented in the way logical thought is, the former is unable to account for the latter. That it has often been the case in our species' history that 2 and 3 have made 5 does not explain our insight that this is necessarily so. And even if arguing in ways that happen to be logical has helped us survive, that does not explain why they are logical or how we know that they are. No matter how many of my ancestors have noticed that if $A = B$ and $B = C$ then $A = C$, that does not explain how I know this is necessarily true. As Lewis argued, reason has a necessity and order that cannot be reduced to the mere regularities of nature. Logic tells us what we ought to think, not what we do think or what our ancestors thought. Deriving logic from nature is just as much a naturalistic fallacy as supposing that nature secretes ethics.[77] The Christian view is that reason is prior to nature, making both it and our thinking orderly enough that we can understand nature (though this ability is now impaired by the Fall). Thus, a thoroughgoing reductionist about reason cannot be a Christian.

Ruse knows there is a problem here, and it resurfaces when he discusses Polkinghorne's argument that selection cannot explain insights of pure mathematics, because pure mathematics never promoted survival. Ruse eventually stipulates that mathematics has a kind of autonomy, but he responds, "I really do not see why a Darwinian should not hold to the Platonic vision as much as a Christian."[78] If Ruse means that a Darwinian such as Kenneth Miller could be a Platonist, he is right. Yet Platonism is surely incompatible with Ruse's version of Darwinism. First, it violates methodological naturalism because numbers, sets, and so forth, are not natural objects, having no location in space and time; yet they are ineliminable from science. But Platonism also violates reductionism, since Platonic categories of eternal and necessary truth cannot be reduced to the contingency or even nomic regularity of nature. Although one might trivially redefine nature to include them, Platonic realities should count as supernatural since they are not governed by natural law ($2 + 3 = 5$ does not suffer from entropy). It follows that if mathematical thought is the interaction of the Platonic with the human mind, it is miraculous. Here is a full-blooded miracle Ruse may have to accept.[79]

Genes and Action

Ruse encounters these problems because, wanting to ride the horse of Darwinism so far, he saddles it with too much baggage. The original question was "Can a Darwinian be a Christian?" Consequently, Ruse's discussion of sociobiology seems gratuitous. Ruse claims that "sociobiology has come into its own as a full member of the Darwinian areas of scientific enquiry."[80] Yet Darwinists of the stature of Stephen Jay Gould,[81] Richard Lewontin,[82] and Kenneth Miller[83] all take a dim view of sociobiology. Once again, it is reductionism, not science, that keeps the horse going.

Sociobiology is the attempt to extend the selection mechanism to human thought, behavior, and culture. For awhile it gratified the physics envy of some social scientists. Then it received such a drubbing from both traditional social scientists and biologists that many of its practitioners are hidden in departments of "anthropology" and "evolutionary psychology."

One problem with sociobiology is that it offers explanations when none are necessary. It has been claimed that because of an altruism gene, old Eskimos stay behind to die when food is short. But as Gould has pointed out, the Eskimos do not need such a gene since they can reason it out for themselves. In fact, the autonomy of reason is fatal to sociobiological claims: "Once reason is admitted as a characteristic of human nature . . . it can be shown to do the work imputed to phantom genes in almost any examples that sociobiologists want to bring up."[84] And as we have seen, Christianity does a far better job than reductionism of supporting the legitimacy and autonomy of reason.

Again, sociobiology derives its prestige as "science" from the claim that it offers a mechanistic explanation of the patterns of human behavior in terms of genetic distribution. But as Lewontin points out, sociobiologists do not say which real genes are responsible, so they do not really provide a mechanism; in addition, they allow so many qualifications to their theories that the theories become vacuous. Lewontin chides:

> The trouble is that if there are behavior genes in humans that affect only some unspecified proportion of the carriers (incomplete penetrance) and with an unspecified variation in the nature of the effect (variable expressivity), no geneticist can confirm their existence.[85]

Part of the problem is that nothing is incompatible with such a weak and vague claim. And too often the sociobiologist offers only "just-so" stories, redescribing what we already know in "scientific language": some people are sometimes patriotic because they carry a hypothetical gene that causes patriotism given some stimuli.[86] Appealing to unidentifiable genes is no better than the pseudoscience ridiculed by Molière: opium induces sleep because of a dormative virtue. Unless I see the gene and put my fingers in its bases, I will not believe.

The only reason I can see for believing such weak science is that it has to be true if reductionism is true. But reductionism is vastly underdetermined by the evidence and seems to play the role of a naturalistic religion. Sociobiology provides comforting narratives to give life an ersatz meaning. Like the astrologist's "The stars herald war," the sociobiologist's "The genes made me do it" gives a sense of pattern and order that we all crave. For reasons already given, it gives no hope that the order is valid.

Reductionism, as understood by scientific materialism, cannot adequately account for miracles, reason, or human action. Reductionism is not only incompatible with Christianity, but it is false. Can a Darwinian be a Christian? By grace anyone can be a Christian. But the camel of Darwinism (it evolved from the horse to cope with the heat of argument) must jettison its reductionism if it is to enter the Christian worldview.

Science and Christianity in Dialogue

Is there a better model for engagement between science and Christianity? I think so, although crude and monolithic thinking about Christianity and science has obscured its existence.

The most obvious weakness in many of the science and religion discussions today is that they are grounded in a simplistic reaction against the warfare of the past. Conflict models had led to the separation of theology and science. The indisputable success of science with the defensive, reactionary character of fundamentalism very properly led to calls for a "humility" theology. Such a theology listens to science and reflects thoughtfully on its findings. Unfortunately, this attractively progressive picture overlooks two important points.

First, if theology alone is called to be humble, then its only available response is accommodation. That is, the job of theologians is simply to find religious significance in whatever the best science is saying. This position assumes without argument that science has a superior status to theology—it is scientific thought that initiates theological reflection, but not vice versa—and that science may not also benefit from humility. The rise of scientism, with its implausible reductionist claims about human thought, meaning, and ethics, shows that science (or what claims to be science) is quite capable of arrogant imperialism in the territories of theology and philosophy. Thus, it is not only that science needs to be humbled but also that theology can sometimes be justified in standing firm. A real dialogue can have agreement and deference as well as disagreement and debate. Humility theology without humility science emasculates true dialogue and produces a kind of "scientific correctness" that stifles free expression in favor of certain approved ways of speaking about science. As a result, scientific ideas whose theological presuppositions do not accommodate to mainstream science—because, for example, they reject methodological naturalism—are excluded from the conversation. This is exactly what happens when the objections of intelligent design to neo-Darwinism are dismissed as nonscience.

Second, the idea that accommodation is the only theological alternative to conflict is naïve. As H. Richard Niebuhr argues in his classic book *Christ and Culture*,[87] three other models are available for interchange between the sacred and the secular. Taking science as an example of secular activity, all of these models fully respect both poles of the dialectic rather than make either Christian theology or science recessive.

For those in the great Thomist tradition of Roman Catholicism, synthesis is the natural model: the best science and the best theology are drawn together in a coherent unity. This is the foundation for various theories of theistic or guided evolution. While the synthesis model is to be commended for taking science and theology in full-strength versions, the danger is that the dominant scientific paradigm may turn out false or incomplete. At one time the Newtonian paradigm was so successful that theology was recast in its image, and philosophers thought they had discovered the necessary structure of human experience, pretensions demolished by relativity theory and quantum mechanics. A kind of Hegelian fallacy rests in supposing that one's own age is the one that has found the final synthesis, exemplified today by scientists such as E. O. Wilson, who makes the grandiose claim that evolution is a spectacular example of the "consilience" of many independent inductive theories and thus an all-encompassing theory of everything.[88]

The Reformed tradition emphasizes the fallenness of human reason and argues that while science is a valuable source of understanding, it is liable to misdirection and so must be transformed before it can be integrated with theology. The transformation model argues that authentically Christian science is different from purely secular science. The approach has considerable merit, raising important questions about the proper direction of science, such as whether it must proceed on the assumption of methodological naturalism. Yet, just as many citizens (including Christians) feel threatened by "Christian government"; even Christian scientists may feel that "Christian science" does not allow them sufficient autonomy to do their work with intellectual honesty.

Finally, the classical Lutheran approach encourages an ongoing dialogue between science and religion, both at full strength. This view takes as its starting point Christ's prayer in John 17 that His disciples be in the world but not of it. Two Kingdoms theology explains that Christians are simultaneously citizens of the left-hand kingdom of the world, which encompasses earthly morality and secular reason (including science), and of the right-hand kingdom of grace, which encompasses God's self-revelation through the Gospel. Due to human fallibility and ignorance, the two kingdoms will often appear to conflict. Since God is sovereign of both kingdoms, the conflict cannot be ignored in favor of an easy accommodation, synthesis, or transformation. Nor does one have the option of separating oneself from the world like a fundamentalist. Instead, the Christian scientist is called to struggle through the difficulties, like Jacob wrestling with God, hoping to find not perfect but the best available solutions, which may still leave many questions unanswered. The great advantage of this model is that its emphasis on human cognitive limitations allows one to hold both scientific and theological views tentatively.

According to the dialogue model, the best way to find the truth is rather Popperian. Human fallibility means we are not sure what is correct, either scientifically or theologically. So we had better allow room for many competing, conflicting voices (Popper's context of discovery) and then engage in debate and attempted falsification (Popper's context of justification). As Popper argued, even unscientific sources may lead to valuable scientific insights. Kepler's Pythagorean mysticism and Newton's belief in a divine lawgiver were immensely fruitful for science. Attempts to dismiss intelligent design because of the religious motivation of some of its proponents show tremendous ignorance of the history of science, which repeatedly documents the importance of religion in scientific creativity.[89] Following Popper's model, those who think that intelligent design is false should allow that model to be articulated in its most powerful, testable form and then attempt to falsify it. Unfortunately, what is actually happening is that many scientists and theologians are conflating these two contexts and trying to dismiss intelligent design a priori as nonscience. This has, in Bertrand Russell's memorable words, all the advantages of theft over honest toil.

To put matters another way, proponents of a variety of theories who wish to offer evidence in favor of them need to be given their "day"—indeed many days—in the court of scientific discussion and evaluation. To do so is important as a corrective to the idols of the mind, which make scientists dismiss new and unusual claims out of hand. In the seventeenth century, Newton's ideas about gravitation were initially rejected as inherently unscientific because they conflicted with the Cartesian, mechanistic assumption that all causation occurs via mechanical contact. When Newton's ideas were finally accepted as a unifying theory of terrestrial and celestial motions, they in turn were held dogmatically despite the humility of Newton himself. In the eighteenth century, John Harrison's sea clocks, designed to establish the longitude of a ship at sea, were initially dismissed because it was assumed that the Newtonian astronomical paradigm was the only scientific means of solving the problem.

> The admirals and astronomers on the Board of Longitude openly endorsed the heroic lunar distance method, even in its formative stages, as the logical outgrowth of their own life experience with sea and sky. . . . In comparison, John Harrison offered the world a little ticking thing in a box. Preposterous![90]

The delay in accepting Harrison's brilliant innovation left sailors without a reliable means of navigation when the heavens were occluded, leading to costly detours and disastrous wrecks, with a great cost to human life and welfare. Again, in the twentieth century, T. D. Lysenko stubbornly ignored the mounting evidence that his neo-Lamarckian ideas were false, a dogmatic projection of his dialectical materialism, leading to crop failure and starvation. The dogmatism continues unabated today with naïve and unsubstantiated claims about what can be learned from the human genome.[91] These examples illustrate the perennial tendency of science to succumb to a dogmatic, inflexible paradigm and show that the only corrective is to tolerate the irritation of alternative views, some of which may turn out to contain an important kernel of truth that the paradigm conceals or overlooks.

The entitlement to hold and express alternative and unpopular ideas is also a political freedom, as John Stuart Mill argued in *On Liberty*. Phillip Johnson reminds us of Mill's important reasons for maintaining that

> the governing majority should never attempt to prevent a dissenting opinion from being heard, no matter how much they dislike it and no matter how certain they are that it is wrong.[92]

If the majority view, whether in government or in the scientific establishment, is wrong, toleration of dissent increases the odds that their errors will eventually be discovered. But even if the majority view is correct, as it often may be, it is more likely to be seen to be correct if it must defend itself against critics. As Mill says:

> If the opinion is right, [the majority] are deprived of the opportunity for exchanging error for truth; if wrong, they lose, what is almost as great a benefit, the clearer perception and livelier impression of truth provided by its collision with error.[93]

Either way, when dissent is tolerated, society has a better chance of an increased understanding of important truths, and this advantage applies in both science and other areas of civil life.

The consequence for education is clear. Science educators must allow that well-formulated dissent is expressed when they teach controversial subjects such as evolutionary theory. To do so requires dialogue between those who accept the theory and those who reject it so that each can hear the other side accurately and thus encourage each other to develop better arguments in favor of their position. The claim that "religion" must be excluded from the secular classroom is a red herring because the viewpoints of all sides are religious, so there is effectively no way to exclude religion. It is also biased and erroneous to suggest that scientific results that happen to support belief in deity should not be presented because that would constitute an attempt to promote a particular religious perspective. As John Warwick Montgomery has noted:

> Those who argue that [such] scientific evidences . . . should not be presented in secular schools because to do so is nothing more than disguised religious proselytizing only display their own prejudices: if hard science leads to theological conclusions, this in no way alters the facticity of the data or the scientific character of the investigation.[94]

That this is indisputably a bias and not a principled objection to all religion in secular schools is shown by the fact that atheism and agnosticism are both religious, yet

> secularists have never worried about the legal or philosophical implications of teaching scientific data—or even philosophical (e.g., evolutionary) speculations—which allegedly support their own atheistic or agnostic views of the universe.[95]

The tired "myth of neutrality" has been thoroughly exposed as a stratagem for secularists to establish their religion while excluding everyone else's.

Conclusion

The protected status that Darwinism enjoys has encouraged a scholastic approach to science that impedes discovery and evades serious testing. Some even try to suppress rational criticism of Darwinism as if it were a matter of excluding heresy. More pacific individuals, such as Michael Ruse, have made serious and thoughtful attempts to reconcile Darwinism and Christianity. If

I am right, such attempts cannot succeed so long as reductionism is tied to Darwinism because this makes science dominant and Christianity recessive in a way that no one serious about the cognitive claims of orthodox Christianity can (or should) allow. Conversations between science and religion will prosper when science and theology each practice humility and when the context of discovery is opened up to allow a greater diversity of ideas to be discussed and developed. We will always have plenty of opportunity to eliminate mistaken theories in the rigors of critical debate and severe testing.

Notes

1. Francis Bacon, *The New Organon*, ed. Lisa Jardine and Michael Silverthorne (New York: Cambridge University Press, 2000), bk. 1, 63 (p. 52).

2. John Warwick Montgomery, "Science, Theology, and the Miraculous," in *Faith Founded on Fact: Essays in Evidential Apologetics* (Edmonton, Canada: Canadian Institute for Law, Theology, and Public Policy, 2001), 56.

3. Many of the criticisms leveled against dogmatic Darwinism in this chapter also apply to scientific materialism in general. As a purely empirical, working hypothesis, Darwinism is not the main target of this chapter but rather the dogmatic attitude of those who see Darwinism as much more than a hypothesis.

4. With William Dembski, Michael Ruse is coediting the forthcoming *Debating Design: From Darwin to DNA* (Cambridge University Press), which includes essays by proponents of Darwinism, self-organization, theistic evolution, and intelligent design.

5. By endorsing some of Bacon's critical responses to scholasticism in this sense, I by no means commit myself to his naïve positive account of true induction, which has been rightly rejected by most philosophers and historians of science.

6. The material that follows is adapted from my book review of *Icons of Evolution*, "Shattering the Icons of Dogmatic Darwinism," *Human Events*, December 1, 2000, 12, 19.

7. Jonathan Wells, *Icons of Evolution: Science or Myth? Why Much of What We Teach about Evolution Is Wrong* (Washington, D.C.: Regnery, 2000).

8. Francis Bacon, *The New Organon*, ed. L. Jardine and M. Silverthorne (Cambridge: Cambridge University Press, 2000), bk. 1, 44 (p. 42).

9. Bacon, *The New Organon*, bk. 1, 3 (p. 33).

10. Theodosius Dobzhansky, "Nothing in Biology Makes Sense Except in the Light of Evolution," *The American Biology Teacher* 35 (1973): 125–29.

11. Wells, *Icons of Evolution*, 245.

12. Bacon, *The New Organon*, bk. 1, 19 (p. 36).

13. Wells, *Icons of Evolution*, 247.

14. See, for example, the work of Harry Clemmey and Nick Bradham, cited in Wells, *Icons of Evolution*, 18.

15. Bacon, *The New Organon*, bk. 1, 12 (p. 35).
16. See Wells, *Icons of Evolution*, 51–54.
17. Wells discusses the circular reasoning in *Icons of Evolution*, 63–65.
18. Karl Popper, "Darwinism as a Metaphysical Research Programme," in *The Philosophy of Karl Popper*, part I, ed. Paul A. Schilpp (La Salle, IL: Open Court, 1974), 133–43.
19. Wells has a Ph.D. in embryology, and his knowledge of embryos was what tipped him off in that something was amiss with Darwinian claims about ontogeny.
20. Wells, *Icons of Evolution*, 87.
21. Wells, *Icons of Evolution*, 101.
22. This in itself is a doubtful move: Darwinism requires gradual evolution and hence a large number of transitional forms, so a significant number should be available. Certainly, to claim that conditions are always such as not to preserve transitional forms conveniently insulates the theory from testing.
23. Wells, *Icons of Evolution*, 124.
24. Wells, *Icons of Evolution*, 134.
25. The cases of Piltdown man and the severe methodological flaws in Kettlewell's widely accepted experiments with peppered moths illustrate the same excessive credulity.
26. See Wells, *Icons of Evolution*, chs. 7 and 8.
27. Judith Hooper, *Of Moths and Men* (New York: Norton, 2002). Hooper argues that it was Kettlewell's desperate desire to please E. B. Ford, an ambitious biologist on a mission to prove natural selection, that led Kettlewell to go to excessively artificial lengths to control his experiments.
28. See Wells, *Icons of Evolution*, ch. 9
29. Wells calls this the "old icon of horse evolution." See *Icons of Evolution*, 196.
30. Wells calls this the "new icon of horse evolution." See *Icons of Evolution*, 199–201.
31. There is, to be sure, a limited sense in which facts are "self-interpreting," as John Warwick Montgomery has pointed out, for it is facts that ultimately arbitrate between competing theories: "the object of the investigation ultimately decides among the interpretations of it, and so we may say that the object is self-interpreting" (J. W. Montgomery, *Tractatus Logico-Theologicus*, 2.3751). In the case of paleoanthropology, however, the problem is massive underdetermination: there are far too few facts to decide between a plethora of widely different theories that equally fit those facts.
32. Wells, *Icons of Evolution*, 223
33. Constance Holden, quoted in Wells, *Icons of Evolution*, 220. The source is Constance Holden's "The Politics of Anthropology," *Science* 213 (1981): 425–38.
34. This is exactly the same weakness we saw in the proposed solutions to Behe's problem of irreducible complexity. The fact that certain narratives are "plausible" and might be true is taken as actual evidence that they are true, a painfully Gnostic perversion of the true scientist's willingness to be surprised by concrete reality.
35. Wells, *Icons of Evolution*, 225.

36. Jonathan Wells, "Critics Rave over *Icons of Evolution*: A Response to Published Reviews," posted on June 12, 2002, at www.discovery.org/viewDB/index.php3?program=CRSC%20Responses&command=view&id=1180.

37. Richard Dawkins, "Put Your Money on Evolution," *The New York Times* (April 9, 1989), section 7, 35.

38. Dennett, *Darwin's Dangerous Idea*, 520.

39. Wells, "Critics Rave over *Icons of Evolution*: A Response to Published Reviews."

40. See, for example, Michael Behe's "Correspondence with Science Journals: Response to Critics Concerning Peer-review," posted in 2000, at www.discovery.org/viewDB/index.php3?program=CRSC%20Responses&command=view&id=450. See also Jonathan Wells's more recent "Catch 23," posted in 2002, at www.discovery.org/viewDB/index.php3?program=CRSC&command=view&id=1212.

41. G. Easterbrook, "Science and God: A Warming Trend?" *Science* 277 (1997): 890–93, 891.

42. Kenyon was interviewed about his experiences by *Mars Hill Audio* in 1994. See audiocassette volume 7, available at www.marshillaudio.org.

43. Stephen Meyer, "Danger: Indoctrination. A Scopes Trial for the '90s," *The Wall Street Journal*, Op. Ed., December 6, 1993, A14. Available online from the article database at www.discovery.org.

44. For more on this topic see my article "Few Signs of Intelligence: The Saga of Bill Dembski at Baylor," *Touchstone*, May 2001, 54–55.

45. This section is adapted from my "Reductionism: Bane of Christianity and Science," *Philosophia Christi* 4, no. 1 (2002): 173–83.

46. In his book *Possibilities for over One Hundredfold More Spiritual Information: The Humble Approach in Theology and Science* (London: Templeton Foundation Press, 2000), Sir John Templeton says that humility theology is open to "the concept that every discovery in any science helps humans to enlarge their definition of the word god" (vii). I would argue that we also need to ask whether theology can enlarge our concept of science.

47. Yet not really science but scientism, the worship of science as the only valid path to objective knowledge.

48. I attempt to lay out a more evenhanded approach to dialogue between science and religion in my "Lutheran Theology Meets Intelligent Design," *Dialog: A Journal of Theology* 40, no. 1 (spring 2001): 61–63.

49. Michael Ruse, *Can a Darwinian Be a Christian? The Relationship between Science and Religion* (New York: Cambridge University Press, 2001).

50. Synopsis of Ruse, *Can a Darwinian Be a Christian?* (page preceding title page, not numbered).

51. Ruse, *Can a Darwinian Be a Christian?* 48.

52. For example, Ruse's discussion of the problem of evil is excellent, both humble and humbling.

53. For effective criticisms of much of the standard evidence for evolution cited by Ruse (13–18), see Michael Denton, *Evolution: A Theory in Crisis* (Chevy Chase, Md.: Adler and Adler, 1986); and Wells, *Icons of Evolution*. Denton, a molecular biologist, shows the weakness of the argument from the fossil record, which gives much stronger evidence of biological discontinuity than Darwinism allows; and the argument from homology to common descent, which is undermined by the frequency of independent convergence to similar adaptations and the overwhelming evidence that the constraints on adaptations are much tighter than mutation–selection allows, an observation Ruse himself discusses (86). Wells gives further support to Denton and thoroughly refutes the argument from embryology, which claims that the similarity in the embryos of organisms of different species is evidence of a common origin. This is simply false. As Wells argues (*Icons of Evolution*, 97–99), the embryos are different right from the beginning. Wells, with a Ph.D. in embryology, should know. To his credit, Ruse does later concede that the biogenetic law (that ontogeny, the development of the individual, recapitulates phylogeny, the evolutionary history of the species) has many counterexamples (25); but, then, why the false statement on page 18? "Organisms of different species with adults of different forms have embryos which are identical."

54. Kenneth R. Miller, *Finding Darwin's God: A Scientist's Search for Common Ground between God and Evolution* (New York: HarperCollins, 1999). I do not agree with all of Miller's positions and would contest some of his evidence for Darwinism. But I do think he is right to carefully distinguish the scientific issue from other philosophical issues, and I appreciate his critique of reductionism and scientism (see chapters 6 and 7 of Miller's book).

55. Ruse, *Can a Darwinian Be a Christian?* 77–80.

56. Ruse, *Can a Darwinian Be a Christian?* 217.

57. If "miracle" means "sign revelatory of God's nature and purposes," it is certainly possible that some miracles are preordained through the normal working of the laws of nature—for example, a rainbow might have a natural explanation and still serve as a promise.

58. Ruse, *Can a Darwinian Be a Christian?* 97.

59. Ruse seems unaware that Christian apologists have convincingly shown Hume's arguments against miracles to be embarrassingly bad. See, for example, William Lane Craig, *Reasonable Faith: Christian Truth and Apologetics*, rev. ed. (Wheaton, IL: Crossway, 1994), 150–54.

60. Ruse, *Can a Darwinian Be a Christian?* 96. Ruse nowhere deals with the obvious objection that eleven of the twelve disciples went to their deaths proclaiming the resurrection as fact. People may die for a lie they believe is true—witness the tragic suicides of the brainwashed kamikaze pilots—but would they die for something they know to be a lie?

61. God became the man Jesus, but he did not cease to be God. For Jesus to fulfill the law perfectly by his life and to pay for all our transgressions by his death, Jesus had to be God.

62. It is not, as Ruse's talk of "reversals" suggests, resuscitation. For a strong defense of the resurrection against naturalistic debunking, see William Lane Craig, "The Resurrection of Jesus," in *Reasonable Faith*.

63. See C. S. Lewis, "Miracles and the Laws of Nature," in *Miracles: A Preliminary Study*, 2d ed. (New York: Macmillan, 1960); and William Lane Craig, "The Problem of Miracles," in *Reasonable Faith*, 142–44. Also helpful are Douglas Geivett and Gary Habermas's collection, *In Defense of Miracles: A Comprehensive Case for God's Action in History* (Downers Grove, IL: IVP, 1997); and Del Ratzsch's *Nature, Design and Science: The Status of Design in Natural Science* (Albany: State University of New York Press, 2001), 120–22.

64. Just how strong a notion of necessity is required is debated: see the discussion in David Armstrong's *What Is a Law of Nature?* (Cambridge: Cambridge University Press, 1983), ch. 11. But my arguments do not depend on how this issue gets resolved. No matter how strong the connection between *A* and *B*, this is consistent with *A*'s being supernaturally changed to *A** and producing *B** instead.

65. In a private correspondence, Elliott Sober objected that if miracles are consistent with laws of nature, then no one should deny them. But those who seek to explain all events in terms of the original resources of the universe will still want to deny them because they cannot be predicted on the basis of those resources and the laws alone. Additional supernaturally added resources mean that the laws produce outcomes other than those we would expect without those resources. In that case, it would be entirely natural for someone to think that a law was broken because, unaware of the supernaturally added element, they misidentify the law in question.

66. It is hardly likely that this can be proved empirically. To do so, one would need to account for every event in the universe by means of its original natural resources. But even our best scientific theories are underdetermined by the evidence so that it is not possible to exclude rival theories, including ones that appeal to supernaturally added resources. But in any case, we have seen that miracles don't violate laws of nature, so evidence for those laws is not evidence against miracles.

67. Humility science would suggest that we might not know what the laws of nature are since we have been so often wrong in the past. Is not the deist's critique of God's miracles like the schoolboy's criticizing Shakespeare for violating the canons of rhyme and meter? Masters of their craft often show that a higher artistic unity can be attained by deviating from the rules taught to an apprentice. The apprentice who criticizes the master for tinkering is imposing parochial and approximate rules of thumb on a higher artistic law. See Lewis, *Miracles*, 96.

68. It is worth objecting to deism that the laws of nature were made for and by God, not God for and by the laws of nature. Of course, one can find "modernist Christians" who hold a deistic view. Just as a rubber duck is no kind of duck, modernist Christianity is no kind of Christianity since it denies the miracles absolutely essential to the faith.

69. See C. S. Lewis, "The Cardinal Difficulty of Naturalism," in *Miracles*. This theme is also in Lewis's *The Abolition of Man* (New York: Macmillan, 1955), where he

argues against reductionist science: "You cannot go on 'explaining away' for ever: you will find that you have explained explanation itself away. You cannot go on 'seeing through' things forever. . . . To 'see through' all things is the same as not to see" (91).

70. Alvin Plantinga, *Warrant and Proper Function* (New York: Oxford University Press, 1993), ch. 12. Plantinga's worry is hardly theoretical. The suggestion of Pinker and Dawkins that our bodies are automata driven by our genes makes human rationality redundant to survival. If correct, there is no ground for supposing reason is reliable and hence no ground for believing Pinker's and Dawkins's arguments. For a critique of these views, see Phillip Johnson's "Darwinism of the Mind," in his *The Wedge of Truth* (Downer's Grove, IL: IVP, 2000).

71. Ruse, *Can a Darwinian Be a Christian?* 107–8.

72. Though I am defending Plantinga's general approach, my arguments are somewhat different, inspired more closely by C. S. Lewis. Despite the similarity, Plantinga says he had not been thinking of Lewis's *Miracles* when formulating his argument.

73. Lewis, *Miracles*, 19.

74. If desired, the random-number generator can even simulate mutation in the rules, but this is not a good idea in practice. Studies of artificial mutation confirm the intuition that random change in complex systems leads to extinction in short order, which is why learning algorithms actually embody intelligent design.

75. Notice that this is exactly what we are told to expect on the account of Pinker and Dawkins. Selection produces a highly adapted robot slave for the preservation and reproduction of its genes.

76. This is a fair characterization of the reductionists' position since they claim that the blind watchmaker can produce human agents who intentionally design and build watches.

77. Even if evolution could explain the origin of some moral sentiments, it cannot explain their accuracy or inaccuracy because the truth-making values are not derivable from natural facts.

78. Ruse, *Can a Darwinian Be a Christian?* 124.

79. To be sure, this would be a miracle only in the philosopher's sense of something supernaturally caused, not in the theologian's sense of a sign revelatory of God's nature and purpose.

80. Ruse, *Can a Darwinian Be a Christian?* 189.

81. Gould's criticism has become muted because, as Tom Bethell points out, "He surely saw the danger—that an attack on sociobiology could damage Darwinism itself" because critics would, as Phillip Johnson said, burn down "the Darwinist house to roast the sociobiological pig." Bethell, "Against Sociobiology," *First Things*, January, 2001, 23.

82. Richard Lewontin, Steven Rose, and Leon Kamin, *Not in Our Genes: Biology, Ideology, and Human Nature* (New York: Pantheon Books, 1984).

83. Ken Miller, *Finding Darwin's God*, 182–91.

84. Tom Bethell, "Against Sociobiology," 22. Of course, it may be objected that genes explain human reasoning. But first, I have already argued that mechanistic ex-

planations do not adequately account for human reason; and second, even if they did, reason is so plastic (it can rationalize anything, and it does so on the basis of experience) that genes could be used to explain virtually any behavior and its opposite, thereby making their invocation vacuous. What we want is an account of Jim's patriotism, not a theory that could equally explain his patriotism and his flag burning.

85. Lewontin, *Not in Our Genes*, 252.

86. Since no means of identifying the gene is given, this really just says some people are sometimes patriotic, which we knew already. Bayes's theorem explains why our experience cannot even add positive confirmation to such claims: they are already to be expected given our background knowledge and yield no interesting testable predictions.

87. H. Richard Niebuhr, *Christ and Culture* (New York: Harper & Row, 1951). For a systematic reappraisal of that work, from a Lutheran point of view, see my collection of essays, *Christ and Culture in Dialogue: Constructive Themes and Practical Applications* (St. Louis, Mo.: Concordia Academic Press, 1999).

88. Edward O. Wilson, *Consilience: The Unity of Knowledge* (New York: Knopf, 1998).

89. See, for example, Peter Harrison, *The Bible, Protestantism, and the Rise of Natural Science* (Cambridge: Cambridge University Press, 1998); Stanley Jaki, *The Savior of Science* (Grand Rapids, Mich.: Eerdmans, 2000); Margaret Osler, ed., *Rethinking the Scientific Revolution* (New York: Cambridge University Press, 2000); and Nancy Pearcey and Charles Thaxton, *The Soul of Science: Christian Faith and Natural Philosophy* (Wheaton, Ill.: Crossway Books, 1994).

90. Dava Sobel, *Longitude: The True Story of a Lone Genius Who Solved the Greatest Scientific Problem of His Time* (New York: Walker Publishing Company, 1995), 98–99.

91. See Richard Lewontin, *It Ain't Necessarily So: The Dream of the Human Genome and Other Projects* (New York: New York Review of Books, 2000); and Barry Commoner's "Unraveling the DNA Myth: The Spurious Foundation of Genetic Engineering," *Harper's Magazine*, February 2002, available at www.mindfully.org/GE/GE4/DNA-Myth-CommonerFeb02.htm.

92. Phillip E. Johnson, *The Right Questions: Truth, Meaning and Public Debate* (Downers Grove, Ill.: Intervarsity Press, 2002), 160.

93. John Stuart Mill, *On Liberty* (Indianapolis, IN: Hackett Publishing, 1978), 16.

94. Montgomery, *Tractatus Logico-Theologicus*, 3.865, 123.

95. Montgomery, *Tractatus Logico-Theologicus*, 3.8651, 124.

~

Index

~

About the Author

A philosopher and a computer science teacher, **Angus Menuge** is associate professor of philosophy and director of the Cranach Institute at Concordia University, Wisconsin (www.cranach.org). Dr. Menuge has done research in the areas of philosophy of mind; history and philosophy of science; Christianity and culture; and Christian apologetics. He is editor of three books: *C. S. Lewis: Lightbearer in the Shadowlands*; *Christ and Culture in Dialogue*; and *Reading God's World: The Vocation of Scientist*. He holds a B.A. in philosophy from the University of Warwick; an M.A. and a Ph.D. in philosophy from the University of Wisconsin, Madison; and a D.C.A. in Christian apologetics from the International Academy of Apologetics, Evangelism, and Human Rights. The present volume synthesizes Menuge's thinking about philosophy of mind with his interests in intelligent design and the philosophy of science.